Chemicals Controlling
Insect Behavior

Contributors

Morton Beroza   J. P. Minyard

Murray S. Blum   J. E. Percy

Nathan Green   Robert M. Silverstein

R. C. Gueldner   A. C. Thompson

D. D. Hardee   H. Harold Toba

Charles Harding   J. H. Tumlinson

P. A. Hedin   David Warthen

Martin Jacobson   J. Weatherston

Chemicals Controlling Insect Behavior

EDITED BY

MORTON BEROZA

Entomology Research Division
Agricultural Research Service

U.S.D.A.
Beltsville, Maryland

Foreword by

F.F. KNIPLING
Entomology Research Division
Agricultural Research Service
U.S.D.A.
Beltsville, Maryland

Academic Press   New York and London 1970

ACADEMIC PRESS, INC.
111 Fifth Avenue, New York, New York 10003

United Kingdom Edition published by
ACADEMIC PRESS, INC. (LONDON) LTD.
Berkeley Square House, London W1X 6BA

LIBRARY OF CONGRESS CATALOG CARD NUMBER: 76-107566

Second Printing, 1972

PRINTED IN THE UNITED STATES OF AMERICA

Contents

Morton Beroza

*Martin Jacobson, Nathan Green, David Warthen, Charles
Harding, and H. Harold Toba*

Robert M. Silverstein

*J. H. Tumlinson, R. C. Gueldner, D. D. Hardee, A. C. Thompson,
P. A. Hedin, and J. P. Minyard*

Murray S. Blum

J. Weatherston and J. E. Percy

Morton Beroza

v

List of Contributors

Numbers in parentheses indicate the pages on which the authors' contributions begin.

Morton Beroza, Entomology Research Division, Argicultural Research Service, U.S.D.A. Beltsville, Maryland (1, 145)

Murray S. Blum, Department of Entomology, University of Georgia, Athens, Georgia (61)

Nathan Green, Entomology Research Division, Agricultural Research Service, U.S.D.A., Beltsville, Maryland (3)

R. C. Gueldner, Entomology Research Division, Agricultural Research Service, U.S.D.A., State College, Mississippi (41)

D. D. Hardee, Entomology Research Division, Agricultural Research Service, U.S.D.A., State College, Mississippi (41)

Charles Harding, Entomology Research Division, Agricultural Research Service, U.S.D.A., Beltsville, Maryland (3)

P. A. Hedin, Entomology Research Division, Agricultural Research Service, U.S.D.A., State College, Mississippi (41)

Martin Jacobson, Entomology Research Division, Agricultural Research Service, U.S.D.A., Beltsville, Maryland (3)

J. P. Minyard, Entomology Research Division, Agricultural Research Service, U.S.D.A., State College, Mississippi (41)

J. E. Percy, Insect Pathology Research Institute, Canada Department of Fisheries and Forestry, Sault Ste. Marie, Ontario, Canada (95)

Robert M. Silverstein,* Life Sciences Research, Stanford Research Institute, Menlo Park, California (21)

A. C. Thompson, Entomology Research Division, Agricultural Research Service, U.S.D.A., State College, Mississippi (41)

H. Harold Toba, Entomology Research Division, Agricultural Research Service, U.S.D.A., Riverside, California (3)

J. H. Tumlinson,† Entomology Research Division, Agricultural Research Service, U.S.D.A., State College, Mississippi (41)

*Present address: Department of Chemistry, State University College of Forestry at Syracuse University, Syracuse, New York
†Present address: Department of Chemistry, State University College of Forestry at Syracuse University, Syracuse, New York

vii

David Warthen, Entomology Research Division, Agricultural Research Service, U.S.D.A., Beltsville, Maryland (3)

J. Weatherston, Insect Pathology Research Institute, Canada Department of Fisheries and Forestry, Sault Ste. Marie, Ontario, Canada (95)

Foreword

As the result of increased interest and research on the biology and physiology of insects, scientists are beginning to recognize the vital role that chemical attractants and repellents play in the behavior of insects. Pheromones produced by insects provide a necessary means of communication in the reproductive processes of many insects. They also seem to be the key to the well-coordinated behavior patterns characteristic of the social insects. Chemists and biochemists are making rapid advances in the techniques of isolation, identification, and synthesis of these highly active and naturally occurring compounds. Structural elucidation of the natural products opens the way for a new and important field of organic chemistry and permits the exploration of new, potentially useful chemicals for controlling insect pests.

These developments are of great interest because the need for effective, economical, and safer means of insect control is more urgent than ever in the past. The expanding world population is creating a greater demand for more food, fiber, and lumber, and this increased production must be met on diminishing land available for agricultural uses. People the world over are also demanding greater protection from the hoards of insects that undermine their health and comfort. Fortunately, scientists have made remarkable progress in dealing with the multitudes of insects that compete for man's food and which threaten his health, even though insects still take a high toll of our agricultural production and insect-borne diseases persist as a major threat to the health of hundreds of millions of people in various parts of the globe. Available methods of insect control depend largely on the use of chemical insecticides having broad-spectrum biological activity. Their use in many situations is not acceptable because of the side effects they produce to nontarget organisms in the environment. Also, insects have a remarkable capability of developing resistance to some of the currently used insect control chemicals. Thus, there is an urgent need for alternative ways to control insects, ways that assure more dependable and permanent effects and yet are not deleterious to other values.

The use of insect attractants and repellents highly selective in action against destructive insects can play an important role in achieving the long-range goals of managing insect populations effectively, economically, and with complete safety to man and his environment. Substantial progress has already been made on chemicals that influence insect behavior, as is evidenced by information contained in this publication. This book brings up-to-date much of the available information on insect pheromones, insect defense mechanisms, and on other insect attractants and repellents. It should lay the groundwork for even greater and more rapid progress in the years ahead. An understanding of the nature of the chemicals that regulate or influence insect behavior is a prerequisite for detailed biological investigations on their potential value in the management of insect populations and in their use for practical insect control.

E. F. KNIPLING

Preface

The increasing use of insecticides and the findings that many of these materials and their metabolites are permeating and contaminating our air, food, water, soil, wildlife, and even our bodies have aroused the concern of a broad spectrum of our community—scientists, public health officials, politicians, and laymen alike—and have engendered a thrust toward exploring noninsecticidal means of insect control. Although this approach is not new, its practical value and present status, as well as the technology involved, are not generally understood or appreciated. The symposium proceedings presented in this volume provide such information on chemicals for controlling insect behavior; such chemicals may prove useful either for direct control or for making other control measures more efficient.

Although certain parts of the book may interest the layman, the contents are directed mainly to the chemist, biochemist, biologist, entomologist, and others working to control insect pests. This broad interest reflects the fact that success has invariably been the product of an interdisciplinary approach. However, interest in the book will not necessarily be limited to those dealing with insects since the means of isolating and identifying microgram and sometimes submicrogram amounts of compounds, such as sex attractants and other pheromones, and biological and behavioral studies with these chemicals are likely to intrigue scientists dealing with natural products and those in the life sciences generally. The scope and treatment of the subject, which must be considered relatively new, are deliberately exemplary rather than exhaustive, with emphasis placed on current trends and practices.

While the safety aspects of the use of nontoxic behavior-controlling chemicals are readily apparent, realistic appraisal of the potential utility of these chemicals for insect control requires adequate background information on structures, modes of action, means of isolation or synthesis, synthetic alternatives, bioassays, and results of field studies. The accomplishments detailed in this work leave little doubt that the use of modern instrumentation and methodology and the rapidly developing expertise of our scientists will bring rapid progress in this area. The elucidation of the chemical structure of minute amounts of insect secretions, their syntheses, the demonstration of their fantastic potency, and the clarification of the biological mechanisms involved are exciting research exploits laden with potential for pest control. It is clear now that many insects depend on chemicals for survival—for finding a mate, for defending themselves, for maintaining their social organization, for finding food, and for appropriate place-ment of eggs. Clearly, the key to insect control in many instances may very well be the key that unlocks the structure of their secretions or of chemicals that attract or repel them.

The papers published in this volume were originally presented at the Symposium on Chemicals Controlling Insect Behavior at the 157th National Meeting of the American Chemical Society in Minneapolis, Minnesota on April 16, 1969.

The excellent assistance of Dr. May N. Inscoe, U.S.D.A., Beltsville, in the editing and indexing of this volume is gratefully acknowledged.

MORTON BEROZA

Chemicals Controlling
Insect Behavior

INTRODUCTION

Morton Beroza

The development of synthetic insecticides during the past several decades stands as one of the great achievements of our times. The saving of millions of lives from insect-borne diseases and the high uninterrupted production of food and fiber from our agricultural community attest to the tremendous value of these chemicals (Whitten, 1966).

Unfortunately, problems relating to the use of insecticides have arisen; some are serious and are causing considerable concern in scientific as well as in lay circles. The problem of resistance to insecticides has been cited frequently (Crow, 1966), and recently we have been alerted to unforeseen disruptive effects of insecticide residues on nontarget ecological systems, systems that were not generally considered in the development of the synthetic insecticides (Barrett, 1968; Hunt, 1966; Moore, 1967; Newsom, 1967). Considering the appearance of residues in our foods, humans may be included as part of one of these nontarget ecosystems. In any event, we are learning more and more about these problems, and a concerted effort has been underway for some time to understand and eliminate these problems as they appear.

Responsible authorities agree that sufficient food to support the world's population cannot be produced without insecticides. With few exceptions, attempts employing biological control alone have not been successful. The same authorities also recognize that new and fresh approaches to pest control must be formulated to minimize pesticide residues, to make the use of pesticides more selective in their attack, and to utilize all means available, including biological control, to overcome insect depredation in a safe manner (Chant, 1966).

As a possible aid in solving some of these problems, we asked leading scientists to discuss chemicals controlling insect behavior. It is fairly certain now that outright killing is not the only way, or even the best way to get rid of our insect competitors. We are seeking more subtle and perhaps more effective means of causing their demise, and — in fact — their eradication if this objective appears feasible. That is why

1

we are interested in disrupting insect growth processes with the insect's own hormones, or in controlling insect behavior patterns on which insects depend for survival. As will become evident in the course of the presentations, investigations along these avenues are not as straightforward as those dealing with the development of insecticides; but we can have confidence that some of the alternative approaches to pest control will in the future prove most useful. We may very well be entering a new era of scientific pest control that will embrace approaches more sophisticated than any used in the past (Knipling, 1966; Smith, 1966).

REFERENCES

Barrett, G. W. (1968). The effects of an acute insecticide stress on a semi-enclosed grassland ecosystem. *Ecology* **49,** 1019.

Chant, D. A. (1966). Integrated control systems. *In* "Scientific Aspects of Pest Control." *Natl. Acad. Sci.-Natl. Res. Council, Publ.* **1402,** 193-218.

Crow, J. F. (1966). Evolution of resistance in hosts and pests. *In* "Scientific Aspects of Pest Control." *Natl. Acad. Sci.-Natl. Res. Council, Publ.* **1402,** 263-275.

Hunt, E. G. (1966). Biological magnification of pesticides. *In* "Scientific Aspects of Pest Control." *Natl. Acad. Sci.-Natl. Res. Council, Publ.* **1402,** 251-262.

Knipling, E. F. (1966). New horizons and the outlook for pest control. *In* "Scientific Aspects of Pest Control." *Natl. Acad. Sci.-Natl. Res. Council, Publ.* **1402,** 455-470.

Moore, N. W. (1967). Effects of pesticides on wildlife. *Proc. Roy. Soc. (London)* **B 167,** 128.

Newsom, L. D. (1967). Consequences of insecticide use on nontarget organisms. *Ann. Rev. Entomol.* **12,** 257.

Smith, E. H. (1966). Advances, problems, and the future of insect control. *In* "Scientific Aspects of Pest Control." *Natl. Acad. Sci.-Natl. Res. Council, Publ.* **1402,** 41-72.

Whitten, J. L. (1966). Pest control and human welfare. *In* "Scientific Aspects of Pest Control." *Natl. Acad. Sci.-Natl. Res. Council, Publ.* **1402,** 401-425.

SEX PHEROMONES OF THE LEPIDOPTERA. RECENT PROGRESS AND STRUCTURE—ACTIVITY RELATIONSHIPS

Martin Jacobson, Nathan Green,
David Warthen, Charles Harding,
and H. Harold Toba

I. INTRODUCTION

The term "pheromone" was coined by Karlson and Butenandt (1959) to designate substances secreted by an animal to influence the behavior of other animals of the same species. It comes from the Greek "pherein" (to carry) and "horman" (to excite, stimulate). As applied to insects, the term "sex pheromones" is now commonly used to include the chemical substances produced and released by one sex to attract or excite the opposite sex for mating. Although the sex pheromones released by female insects may attract males from a distance, they may also serve to excite the male sexually before copulation. Sex pheromones released by males are usually for the purpose of sexually exciting the female, making her more receptive to the male's advances; they are thus in the nature of aphrodisiacs.

The comprehensive review of insect sex attractants by Jacobson (1965) lists 159 species in which sex pheromones produced by the female have been demonstrated; of these, 109 species are members of the order Lepidoptera, which includes the moths and butterflies. A total of 53 species, of which 40 are lepidopterous insects, is given in which the pheromone is produced by the male.

3

In 1965, Jacobson (1965) stated "that the insect sex attractant literature has recently grown by leaps and bounds is due in no small measure to the fascinating fact that these substances, produced by the insects themselves as a requisite to their reproduction, may be used for their destruction." He cited 425 references on the subject appearing between 1837 and 1965. The field has seen an even greater growth since then, however; more than 400 additional references on insect sex pheromones have appeared in the literature during the past 4 years.

Detected by the insect in fantastically minute amounts, the sex attractants are undoubtedly among the most potent physiologically active substances known today. Thus far, their main practical use has been in insect survey, catches in traps baited with the attractants indicating the size and location of infestations of destructive insects, so that control measures with insecticidal treatments could be limited to those areas where they might be needed. Recent investigations have shown however, that the sex attractants themselves may also be useful in insect control (Jacobson, 1965, pp. 112-121).

The sex pheromones are detected by the insect by means of sense organs located mainly in the antennae, and in the case of the Lepidoptera the antennae are probably the sole organs of chemoreception since moths deprived of both antennae cannot locate a mate and show no response when exposed to a pheromone. Schneider (1964, 1969) has shown that the sex pheromones are detected by moths through numerous sensory cells located in the antennae, from which he succeeded in recording action potentials (EAG's or electroantennograms).

Good reviews on various phases of insect sex pheromones are those by Wilson (1963), Jacobson and Beroza (1963, 1964), Jacobson (1966b), Butler (1967), Roth (1967), and Regnier and Law (1968).

II. RECENT DEVELOPMENTS

A. SEX PHEROMONE DEMONSTRATION

In the past 4 years the presence of sex pheromones has been demonstrated in 31 additional species of lepidopterous insects (Table I). In each of these cases, the pheromone is produced by the adult female in glands situated in the last few abdominal segments. In three of these species — the false codling moth *(Argyroploce leucotreta)* (Read *et al.,* 1968), the red-banded leaf roller *(Argyrotaenia velutinana)* (Roelofs

TABLE I

Lepidopterous Insects Demonstrated to Possess Sex Pheromone (1965-1969)

Scientific name	Common name	Reference
Acrolepia assectella (Zeller)	Leek moth	Rahn (1966)
Alabama argillacea (Hübner)	Cotton leafworm	Berger (1968)
Anagasta kühniella (Zeller)	Mediterranean flour moth	Traynier (1968)
Ancylis comptana fragariae (Walsh & Riley)	Strawberry leaf roller	Roelofs and Feng (1968)
Archips argyrospilus (Walker)	Fruit tree leaf roller	Roelofs and Feng (1968)
Archips mortuanus Kearfott	—	Roelofs and Feng (1968)
Argyroploce leucotreta Meyr.	False codling moth	Read *et al.* (1968)
Argyrotaenia quadrifasciana (Fernald)	—	Roelofs and Feng (1968)
Argyrotaenia velutinana (Walker)	Red-banded leaf roller	Roelofs (1966)
Autographa biloba (Stephens)	—	Berger and Canerday (1968)
Choristoneura fumiferana (Clemens)	Spruce budworm moth	Findlay and MacDonald (1966)
Choristoneura rosaceana (Harris)	Oblique-banded leaf roller	Roelofs and Feng (1968)
Chrysopeleia ostryaella Chambers	Leaf miner	Lindquist and Bowser (1966)
Crambus trisectus (Walker)	Webworm	Banerjee (1967)
Diparopsis castanea (Hampson)	Red bollworm	Tunstall (1965)
Epiphyas postvittana (Walker)	—	Bartell and Shorey (1969)
Feltia subterranea (F.)	Granulate cutworm	Jefferson *et al.* (1968)
Grapholitha funebrana (Treitschke)	Plum fruit moth	Saringer *et al.* (1968)
Hedya nubiferana (Haworth)	—	Roelofs and Feng (1968)
Heliothis phloxiphagus Grote & Robinson	—	Jefferson *et al.* (1968)
Hemileuca maia (Drury)	Buck moth	Earle (1967)
Pandemis limitata (Robinson)	Three-lined leaf roller	Roelofs and Feng (1968)
Prionoxystus robiniae (Peck)	Carpenterworm	Solomon and Morris (1966)
Pseudoplusia includens (Walker)	—	Shorey *et al.* (1968); Berger and Canerday (1968)
Rachiplusia ou (Guenée)	—	Shorey *et al.* (1968); Berger and Canerday (1968)
Sitotroga cerealella (Olivier)	Angoumois grain moth	Keys and Mills (1968)
Spodoptera frugiperda (J. E. Smith)	Fall armyworm	Sekul and Cox (1965)
Thyridopteryx ephemeraeformis (Haworth)	Evergreen bagworm moth	Kaufmann (1968)
Vitula edmandsae (Packard)	Bumble bee wax moth	Weatherston and Percy (1968)
Zeadiatraea grandiosella (Dyar)	Southwestern corn borer	Davis and Henderson (1967)

and Arn, 1968a,b), and the fall armyworm *(Spodoptera frugiperda)* (Sekul and Sparks, 1967)—the pheromone has been identified (see below).

Male butterflies of the genus *Lycorea* possess a pair of extrusible, odoriferous, brushlike structures on the posterior of their abdomens called "hair pencils." While in aerial pursuit of the female, the male extrudes his "hair pencils" and brushes them against the female's antennae, thus inducing her to alight. The male continues "hair penciling" the female until she is acquiescent and copulation then occurs. Such use of a form of aphrodisiac by the male has been reported by Brower *et al.* (1965) in the queen butterfly *Danaus gilippus berenice* (Cramer), and by Meinwald *et al.* (1966) in *Lycorea ceres ceres* (Cramer), from which 1-hexadecanol (cetyl) acetate, *cis*-11-octadecen-1-ol acetate, and 2,3-dihydro-7-methyl-1*H*-pyrrolizidin-1-one have been isolated (Meinwald and Meinwald, 1966).

A very interesting development has been reported by Riddiford and Williams (1967a,b), who have shown that polyphemus moths *Antherea polyphemus* (Cramer) can normally mate only in the presence of leaves of the red oak *Quercus rubra*. A volatile leaf emanation, identified by Riddiford (1967) as *trans*-2-hexenal, impinges on the sensory receptors of the female's antennae, triggering the release of her sex pheromone (unidentified), which, in turn, is necessary for the sexual activation of the male.

B. *IDENTIFICATION AND SYNTHESIS OF SEX PHEROMONES*

In the past 4 years the sex pheromones of five destructive lepidopterous insects have been isolated, identified, and synthesized.

1. Pink Bollworm Moth, *Pectinophora gossypiella* (Saunders)

By means of column and preparative gas chromatography of a methylene chloride extract of 850,000 virgin female moths, Jones *et al.* (1966) isolated 1.5 mg of the pheromone, which they identified as 10-propyl-*trans*-5,9-tridecadien-1-ol acetate (I).

$$(CH_3CH_2CH_2)_2 C=CH(CH_2)_2 \overset{H}{\underset{H}{C}}=C(CH_2)_4 O\overset{O}{\overset{\|}{C}}CH_3$$

(I)

The material, which they designated "propylure," proved to be highly exciting to males in laboratory cages, but did not attract males to field traps, although the crude female extract was attractive in the field. Extract from which the propylure had been separated however, was not attractive in the field (Table II). Jones and Jacobson (1968) subsequently found that *N,N*-diethyl-*m*-toluamide (Deet), present in the crude extract, activates propylure so that 1:10 mixtures of 50 female equivalents of propylure and Deet will lure males in the field, although catches are not as high as those obtained with the crude extract; it is probable that a second activator may also be necessary in the mixture to give optimum catches. It is interesting to note that Deet, present in large amount in the adult female moth, has never before been reported from a natural source, but it is a very effective mosquito repellent (McCabe *et al.,* 1954). Propylure has been synthesized by Jones *et al.* (1966), Eiter *et al.* (1967), and Pattenden (1968). Although the preparation of Eiter *et al.* (1967) failed to excite caged males in the laboratory, this was due to a masking problem which will be discussed later.

TABLE II

Attractiveness of Propylure to Male Pink Bollworms in the
Laboratory and in Outdoor Cages

Material tested (50 female equiv.)	Laboratory bioassay	Number of males caught (out of 100)		
		Night 1	Night 2	Night 3
Propylure				
Synthetic	Strong	3	2	1
Natural	Strong	0	0	0
Inactive extract (40 mg)	Negative	0	0	0
Propylure + inactives	Strong	52	56	41
Natural extract (standard)	Strong	62	97	52

2. Cabbage Looper Moth, *Trichoplusia ni* (Hubner)

The pheromone, which excites caged males in the laboratory and attracts males in the field, was isolated from a methylene chloride extract of the last two or three abdominal segments of virgin female moths; it was identified as *cis*-7-dodecen-1-ol acetate (II) by Berger (1966).

$$\underset{\text{H H}}{\text{CH}_3(\text{CH}_2)_3}\text{C}=\text{C}(\text{CH}_2)_6\overset{\overset{\displaystyle O}{\parallel}}{\text{OC}}\text{CH}_3$$

(II)

Berger (1966) synthesized the attractant in an overall yield of 22%. A more direct procedure developed by Green *et al.* (1967) gives an overall yield of at least 37% and the method has been adapted to commercial production of the attractant.

Berger and Canerday (1968) reported that the attractant is not specific for cabbage looper males, since it elicits sexual excitement in caged males of *Pseudoplusia includens* (Walker), *Rachiplusia ou* (Guenée), and *Autographa biloba* (Stephens). Extracts prepared from females of each of these species all contained a component with the same gas chromatographic retention time as (II), but this is no assurance that (II) is common to all of these species. Shorey *et al.* (1965) have reported that an extract of female cabbage loopers will also elicit a sexual response in the male alfalfa looper moth, *Autographa californica* (Speyer).

3. Fall Armyworm Moth, *Spodoptera frugiperda* (J. E. Smith)

A powerful male sex excitant was isolated from an ether extract of virgin female moths by Sekul and Sparks (1967), who obtained 900 μg of pure material from 135,000 insects. These investigators identified the pheromone as *cis*-9-tetradecen-l-ol acetate (III).

$$\underset{\text{H H}}{\text{CH}_3(\text{CH}_2)_3}\text{C}=\text{C}(\text{CH}_2)_8\overset{\overset{\displaystyle O}{\parallel}}{\text{OC}}\text{CH}_3$$

(III)

Although the pheromone is a powerful male mating stimulant in the laboratory (Sekul and Cox, 1965, 1967), it does not appear to be a long-range attractant in the field.

The pheromone was synthesized by Sekul and Sparks (1967) by reducing the extremely rare methyl myristoleate with lithium aluminum hydride to myristoleyl alcohol and acetylating this to give (III). On the other hand, the synthetic procedure developed by Warthen (1968) is commercially adaptable and much more economical. The pheromone has also been synthesized (Jacobson and Harding, 1968) by using the cabbage looper sex attractant (II) as starting material.

4. Red-Banded Leaf Roller Moth, *Argyrotaenia velutinana* (Walker)

Preliminary investigations by Roelofs and Feng (1967) showed that ether and methylene chloride extracts of the last two abdominal segments of the virgin female moths were attractive to males in the laboratory. The active material was isolated from a methylene chloride extract by Roelofs and Arn (1968a,b), who obtained 200 μg from the abdomens of 40,000 insects. They identified the pheromone as *cis*-11-tetradecen-1-ol acetate (IV).

$$\underset{\text{H}}{\text{CH}_3\,\text{CH}_2\,\text{C}}=\underset{\text{H}}{\text{C}}(\text{CH}_2)_{10}\,\overset{\overset{\text{O}}{\|}}{\text{O}}\text{C}\,\text{CH}_3$$

(IV)

Roelofs and Arn (1968b) succeeded in synthesizing the pheromone by two different routes and showed that traps baited with it would attract large numbers of male moths in apple orchards.

5. False Codling Moth, *Argyroploce leucotreta* Meyr.

The pheromone of this citrus pest of South Africa was isolated from an extract of the female insects by Read *et al.* (1968) and shown to be identical with *trans*-7-dodecen-1-ol acetate (V) synthesized by Green *et al.* (1967).

$$\text{CH}_3(\text{CH}_2)_3\,\underset{\text{H}}{\overset{\text{H}}{\text{C}}}=\text{C}(\text{CH}_2)_6\,\overset{\overset{\text{O}}{\|}}{\text{O}}\text{C}\,\text{CH}_3$$

(V)

Shown to excite sexually male moths in the laboratory when tested by the method of Shorey *et al.* (1964), the compound is thus the trans form of the cabbage looper pheromone.

III. STRUCTURE–ACTIVITY RELATIONSHIPS

In a paper presented in 1965 before the Medicinal Chemistry Division at the 150th National Meeting of the American Chemical

Society, Jacobson predicted that the sex attractants of many species of insects, especially the Lepidoptera, would prove to be long-chain, unsaturated alcohols or their esters (Anonymous, 1965). At that time the sex pheromones of only three species of Lepidoptera had been identified, namely, *trans*-10,*cis*-12-hexadecadien-l-ol (bombykol) for the silkworm moth [*Bombyx mori* (L.)], (+)-10-acetoxy-*cis*-7-hexadecen-l-ol (gyptol) for the gypsy moth [*Porthetria dispar* (L.)], and 10-propyl-*trans*-5,9-tridecadien-l-ol acetate (propylure) for the pink bollworm moth. The signs from work already in progress were unmistakable, however, and the sex pheromones of the additional lepidopterous species thus far identified have indeed proved to be long-chain, unsaturated acetates. A pattern appears to be unfolding in the love life of this order of insects and, although it is still too early to predict the specific pheromone structure for a given species, even that may be possible before too long.

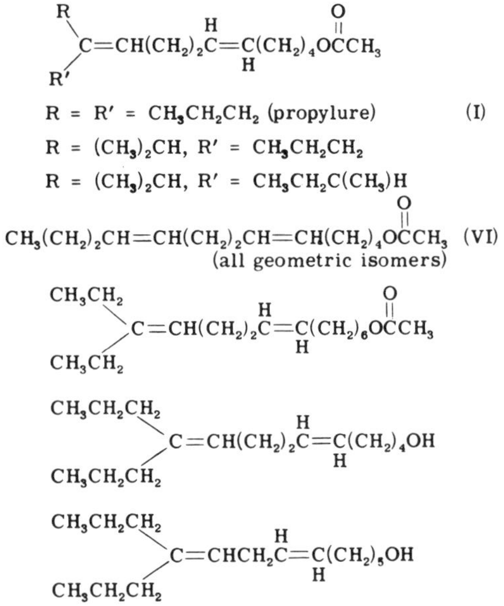

$$\begin{array}{c} R \\ \diagdown \\ \diagup \\ R' \end{array} C=CH(CH_2)_2\overset{H}{\underset{H}{C}}=C(CH_2)_4O\overset{O}{\overset{\|}{C}}CH_3$$

R = R' = $CH_3CH_2CH_2$ (propylure)   (I)

R = $(CH_3)_2CH$, R' = $CH_3CH_2CH_2$

R = $(CH_3)_2CH$, R' = $CH_3CH_2C(CH_3)H$

$$CH_3(CH_2)_2CH=CH(CH_2)_2CH=CH(CH_2)_4O\overset{O}{\overset{\|}{C}}CH_3 \quad (VI)$$
(all geometric isomers)

$$\begin{array}{c} CH_3CH_2 \\ \diagdown \\ \diagup \\ CH_3CH_2 \end{array} C=CH(CH_2)_2\overset{H}{\underset{H}{C}}=C(CH_2)_6O\overset{O}{\overset{\|}{C}}CH_3$$

$$\begin{array}{c} CH_3CH_2CH_2 \\ \diagdown \\ \diagup \\ CH_3CH_2CH_2 \end{array} C=CH(CH_2)_2\overset{H}{\underset{H}{C}}=C(CH_2)_4OH$$

$$\begin{array}{c} CH_3CH_2CH_2 \\ \diagdown \\ \diagup \\ CH_3CH_2CH_2 \end{array} C=CHCH_2\overset{H}{\underset{H}{C}}=C(CH_2)_5OH$$

FIG. 1. Structures of propylure and related compounds synthesized.

In an effort to uncover compounds with attraction for male pink bollworm moths in the field, the compounds whose structures are shown in Fig. 1 were synthesized. None of the compounds shown, with the exception of propylure, elicited sexual excitement in caged pink bollworm males, nor did they attract males in field tests. Screening tests in the laboratory for attraction to cabbage looper (Toba *et al.,* 1968) males showed that the *cis*-5,*cis*-9 and *trans*-5,*cis*-9 isomers of un-branched structure (VI) were attractive to male fall armyworm moths, whereas the *cis*-5,*trans*-9 and *trans*-5,*trans*-9 isomers were not, although the *cis*-5,*trans*-9 acetate caused copulatory behavior (Warthen and Jacobson, 1967). Hypothesizing that the C-5 double bond was unneces-sary for activity whereas the cis configuration in the 9-position was essential, and that the natural fall armyworm pheromone possessed an even number of carbon atoms, since acetates with an odd number of carbon atoms rarely occur in nature, Warthen (1968) prepared the even-numbered 14-carbon acetate having a cis configuration at the C-9 double bond and saturation at the C-5 position. Shortly before the synthesis of this compound was completed, Sekul and Sparks (1967) identified the natural sex pheromone of the female fall armyworm moth as *cis*-9-tetradecen-l-ol acetate (III).

The discovery that the sex pheromones of the pink bollworm, cabbage looper, and fall armyworm moths were all acetates of unsatu-rated 12–16-carbon alcohols prompted us to synthesize a series of congeners for screening with these species of insects (Table III). None of the compounds tested other than the natural pheromone (*cis*-9-tetradecen-l-ol acetate) was active with fall armyworm males, al-though, as was stated previously, the acetates of *cis*-5,*cis*-9- and *trans*-5,*cis*-9-tridecadien-l-ol elicited sexual excitement in caged males. In the case of cabbage looper, very weak activity (compared with *cis*-7-dode-cen-l-ol acetate) was elicited by the acetates of *cis*-5-, *trans*-5-, *cis*-6-, and *trans*-7-dodecen-l-ol as well as the acetate of *cis*-7-tetradecen-l-ol. These data and the work of Jacobson *et al.* (1968) and Berger and Canerday (1968) show that any change in the cabbage looper phero-mone structure results in a pronounced reduction in activity. A considerable number of the compounds in Table III were likewise tested for laboratory attraction to adult males of the codling moth *Laspeyresia pomonella* (L.) and gypsy moth *Porthetria dispar* (L.), both of which are members of the Lepidoptera, and the Mediterranean fruit fly *Ceratitis capitata* (Wiedemann), melon fly *Dacus cucurbitae* Coquillet, and oriental fruit fly *D. dorsalis* Hendel, members of the order Diptera; no activity was observed. Similarly, the acetates of 7-octyn-l-ol, 3-, 6-, 7-, and 8-dodecyn-l-ol, and 5-, 7-, and 8-hexadecyn-l-ol had no effect on these species.

In direct contrast, *cis*-7-hexadecen-l-ol acetate causes high sexual excitement among caged male pink bollworm moths (Table III) and field traps baited with 10–60 mg of this compound alone lure large numbers of males for at least several weeks. A patent application (Green and Keller, 1968) has been filed covering the use of this powerful attractant, to which the name "hexalure" has been given.

TABLE III

Sex Attractant Activity of Various Alkenol Acetates to Males
of the Cabbage Looper, Fall Armyworm, and Pink Bollworm[a]

Acetate of	Activity of		
	Cabbage looper	Fall armyworm	Pink bollworm
Hexen-1-ol			
trans-3-	−	N	−
7-Octen-1-ol	−[b,c]	−	−
Decen-1-ol			
cis-7-	+[c]	−	−
trans-7-	−	−	+
10-Undecen-1-ol	−	−	−
Dodecen-1-ol			
cis-3-	N	N	−
cis-5-	−[d,e]	−[e]	−[e]
trans-5-	−[e]	−[e]	−[e]
cis-6-	+[f]	N	−
trans-6-	+	N	−
cis-7-	++[b,d]	−[b]	−[b]
trans-7-	+[b,d]	−[b]	−[b]
cis-8-	−	N	N
trans-8-	−	N	N
cis-9-	−[d]	N	−
trans-9-	−	N	−
11-	−	N	N
Tetradecen-1-ol			
cis-5-	−	−	−
trans-5-	−	−	−
cis-7-	+[d]	−	N
trans-7-	+	−	−
cis-9-	−	++[g]	−
trans-9-	−	−	−
Hexadecen-1-ol			
trans-2-	−	−	−
cis-7-	+	N	++
trans-7-	+	N	−

TABLE III (cont'd.)

Acetate of	Activity of		
	Cabbage looper	Fall armyworm	Pink bollworm
Hexadecen-1-ol (cont'd.)			
cis-8-	N	N	–
trans-8-	N	N	–
cis-9-	–	N	–
cis-11-	–	N	–
Octadecen-1-ol			
cis-9-	–	–	–

[a]Tested by laboratory bioassay; activity is based on observation of copulatory behavior at the odor source, –, denotes inactive; +, low activity; ++, highly active; N, not tested.

[b]From Green *et al.* (1967).

[c]From Jacobson *et al.* (1968).

[d]Confirmed by Berger and Canerday (1968).

[e]From Warthen and Jacobson (1968).

[f]Berger and Canerday (1968) report this compound to be inactive on cabbage looper males.

[g]From Sekul and Sparks (1967).

As a result of an investigation of the sex pheromone of the female codling moth, McDonough *et al.* (1969b) found that the pheromone is a long-chain, unsaturated alcohol which probably contains an ether group. This prompted them (McDonough *et al.,* 1969a) to synthesize for screening the epoxidation products of fourteen C_8–C_{12} monoolefins, and of geraniol, linalool, farnesol, *cis*-7-dodecen-1-ol, *cis*-9-octadecen-1-ol, and *trans*-9,*trans*-12-octadecadien-1-ol. Of the 21 epoxides prepared, 20 elicited a high degree of sexual excitement in caged male codling moths. The most potent pheromone in this group was *cis*-9,10-epoxyoctadecan-1-ol, which evoked a response at a concentration of only 10^{-9} gm/ml. We prepared the epoxides of *cis*-7-decen-1-ol, *trans*-7-decen-1-ol, *cis*-7-decen-1-ol acetate, *trans*-2-tetradecen-1-ol, *cis*-9-tetradecen-1-ol acetate, *cis*-7-hexadecen-1-ol acetate, and *cis*-9-octadecen-1-ol. Only *cis*-7,8-epoxyhexadecanol acetate and *cis*-9,10-epoxyoctadecanol evoked a sexual response in male codling moths, but the latter also stimulated male gypsy moths sexually.

Butt *et al.* (1968) have recently shown that a number of nitriles, especially heptanenitrile and 5,7,7-trimethyl-2-octenenitrile, sexually stimulate male codling moths in the laboratory and are attractive in the field, although the responses are weak when compared with that of the natural sex attractant. It is of special interest to note that octadecanenitrile is a potent attractant for the male spruce budworm moth *Choristoneura fumiferana* (Clemens) in the field, although it is not the natural sex lure for this insect (Findlay *et al.,* 1967).

IV. POTENTIAL USES OF PHEROMONES FOR INSECT CONTROL

A good review of the theory of this subject with regard to insect orientation and its prevention appeared in 1967 (Shorey and Gaston, 1967), and only the more recent developments will be considered here.

A. FACTORS CAUSING ORIENTATION

Henneberry *et al.* (1967a,b) have shown that catches of male cabbage looper moths with light traps can be increased manyfold by the use of blacklight lamps baited with live virgin females, and Howland *et al.* (1969) reported that such light traps baited each night with 0.5 mg of the synthetic pheromone mixed with sand substrate significantly increased male catches over a period of several weeks. Inasmuch as the attractant is readily available in large amounts and field tests have indicated that it has a marked effect for 400 ft. downwind (Green *et al.,* 1967), the use of the attractant for control of this pest may have excellent potential.

From the standpoints of convenience, stability, long-range effectiveness, and economy, the use of hexalure in field trapping of male pink bollworm moths offers a valuable potential method to aid in the control of this destructive pest of cotton (Green and Keller, 1968).

B. PREVENTION OF ORIENTATION (PHEROMONE MASKING)

The phenomenon of pheromone masking or inhibition has been reviewed recently by Jacobson (1965, 1966a, 1969). It may be subdivided into masking by aerial permeation with pheromone and by chemical competitive action.

1. Aerial Permeation with Pheromone

This technique is based on permeation of the atmosphere with the sex pheromone. Assuming male adaptation to the omnipresent pheromone, and the fact that the additional increment of pheromone contributed by females in nature may be imperceptible to the males, the male would theoretically be incapable of orienting to and inseminating the females. Field trials reported by Gaston *et al.* (1967) showed the cabbage looper sex pheromone concentration to be sufficiently high from 100 stakes each containing 17 mg of synthetic attractant placed at 3-m intervals in a 27-m³ plot to prevent the males from orienting to the pheromone released by living females. Shorey *et al.* (1967) calculated that a concentration of cabbage looper pheromone of roughly 10^{-10} gm/ liter is sufficient to prevent completely the orientation of males to pheromone-emitting females. For large area control, less than 0.2 gm/ acre must be expended per night. These results indicate that economic control of an insect over large areas may be possible by behavioral control with the sex pheromone.

2. Use of Other Chemicals

It has been shown that the activity of a number of pheromones may be masked by other chemicals present in the vicinity of or in admixture with the pheromone (Jacobson, 1969). According to Shorey and Gaston (1967), the mechanisms for such an effect would be adaptation, causing inhibition of responsiveness of the male, or repellency, preventing males from responding to the pheromone.

According to Riddiford and Williams (1967b), the action of *trans*-2-hexenal needed to trigger production of sex attractant by female polyphemus moths can be masked by numerous other volatile agents, including Chanel No. 5. The action of the sex attractant on the male can be completely and reversibly blocked by the vapors of formaldehyde.

Roelofs and Comeau (1968) found that baiting field traps with equal amounts of red-banded leaf roller sex pheromone (*cis*-11-tetradecen-l-ol acetate) (RiBLuRe) and *trans*-11-tetradecen-l-ol acetate, *cis*-11-tetradecen-l-ol, *cis*-9-tetradecen-l-ol, 11-tetradecyn-l-ol, or 11-tetradecyn-l-ol acetate greatly inhibited or even abolished male catch. As little as 5% *trans*-11-tetradecen-l-ol acetate inhibited the response to 10 μg of the pheromone. Tests with a large number of related chemical substances showed that, in general, the trans configuration was more inhibitory than the corresponding cis configuration, and alcohols were more inhibitory than corresponding acetates; inhibitory effects decreased as the position of unsaturation was moved from C-11 to lower odd-numbered carbons.

A sample of propylure (10-propyl-*trans*-5,9-tridecadien-1-ol acetate) prepared by Eiter *et al.* (1967) and found to be inactive on caged male pink bollworm moths was separated by Jacobson (1969) into 60% trans and 40% cis isomers. The investigation disclosed that as little as 10% of the cis isomer admixed with the trans form strongly depressed male response and 15% of the former completely nullified the activity of the latter. Males exposed to the trans isomer within 15 min after exposure to the cis isomer failed to respond, but a complete response was obtained each time in consecutive exposures to the trans isomer alone.

"The possibility of finding compounds (either structurally related or unrelated to natural pheromones) that, by adaptation of pheromone receptors or an effect on orientation mechanisms, prevent male responses to females in nature has great appeal" (Shorey and Gaston, 1967).

ACKNOWLEDGMENTS

We thank the following Entomology Research Division personnel for conducting biological tests with our synthetic compounds: Dr. L. F. Steiner, Honolulu, Hawaii; Dr. M. T. Ouye, Brownsville, Texas; Dr. A. A. Sekul, Tifton, Georgia; and Mr. B. A. Butt, Yakima, Washington. We also thank Mr. F. M. Phillips, Plant Pest Control Division, Otis Air Force Base, Massachusetts for the biological test results with gypsy moth.

REFERENCES

Anonymous. (1965). Many sex attractants may well prove to be long-chain, unsaturated alcohols or esters. *Chem. Eng. News* **43,** (No. 38) 42.

Banerjee, A. C. (1967). Flight activity of the sexes of crambid moths as indicated by light-trap catches. *J. Econ. Entomol.* **60,** 383.

Bartell, R. J., and Shorey, H. H. (1969). A quantitative bioassay for the sex pheromone of *Epiphyas postvittana* (Lepidoptera) and factors limiting male responsiveness. *J. Insect Physiol.* **15,** 33.

Berger, R. S. (1966). Isolation, identification, and synthesis of the sex attractant of the cabbage looper, *Trichoplusia ni. Ann. Entomol. Soc. Am.* **59,** 767.

Berger, R. S. (1968). Sex pheromone of the cotton leafworm. *J. Econ. Entomol.* **61,** 326.

Berger, R. S., and Canerday, T. D. (1968). Specificity of the cabbage looper sex attractant. *J. Econ. Entomol.* **61,** 452.

Brower, L. P., Brower, J. and Cranston, F. P. (1965). Courtship behavior of the queen butterfly, *Danaus gilippus berenice* (Cramer). *Zoologica* **50**, 1.

Butler, C. G. (1967). Insect pheromones. *Biol. Rev. Cambridge Phil. Soc.* **42**, 42.

Butt, B. A., Beroza, M., McGovern, T. P., and Freeman, S. K. (1968). Synthetic chemical sex stimulants for the codling moth. *J. Econ. Entomol.* **61**, 570.

Davis, F. M., and Henderson, C. A. (1967). Attractiveness of virgin female moths of the southwestern corn borer. *J. Econ. Entomol.* **60**, 279.

Earle, N. W. (1967). Demonstration of a sex attractant in the buck moth, *Hemileuca maia* (Lepidoptera: Saturniidae). *Ann. Entomol. Soc. Am.* **60**, 277.

Eiter, K., Truscheit, E., and Boness, M. (1967). Synthesen von D,L-10-Acetoxy-hexadecen-(7-*cis*)-ol-(1), 12-Acetoxy-octadecen-(9-*cis*)-ol-(1) ("Gyplure") und 1-Acetoxy-10-propyl-tridecadien-(5-*trans*,9). *Ann. Chem.* **709**, 29.

Findlay, J. A., and MacDonald, D. R. (1966). Investigation of the sex-attractant of the spruce budworm moth. *Chem. Can.* 1966, 3.

Findlay, J. A., MacDonald, D. R., and Tang, C. S. (1967). A synthetic attractant for the male spruce budworm moth *Choristoneura fumiferana* (Clem.) *Experientia* **23**, 377.

Gaston, L. K., Shorey, H. H., and Saario, C. A. (1967). Insect population control by the use of sex pheromones to inhibit orientation between the sexes. *Nature* **213**, 155.

Green, N., and Keller, J. C. (1968). Attractant for pink bollworm moths. U. S. Patent Appl. 780,596 (Dec. 2).

Green, N., Jacobson, M., Henneberry, T. J., and Kishaba, A. N. (1967). Insect sex attractants. VI. 7-Dodecen-l-ol acetates and congeners. *J. Med. Chem.* **10**, 533.

Henneberry, T. J., Howland, A. F., and Wolf, W. W. (1967a). Combinations of blacklight and virgin females as attractants to cabbage looper moths. *J. Econ. Entomol.* **60**, 152.

Henneberry, T. J., Howland, A. F., and Wolf, W. W. (1967b). Recovery of released male cabbage looper moths in traps equipped with blacklight lamps and baited with virgin females. *J. Econ. Entomol.* **60**, 532.

Howland, A. F., Debolt, J. W., Wolf, W. W., Toba, H. H., Gibson, T., and Kishaba, A. N. (1969). Field and laboratory studies of attraction of the synthetic sex pheromone to male cabbage looper moths. *J. Econ. Entomol.* **62**, 117.

Jacobson, M. (1965). "Insect Sex Attractants." Wiley (Interscience), New York.

Jacobson, M. (1966a). Masking the effects of insect sex attractants. *Science* **154**, 422.

Jacobson, M. (1966b). Chemical insect attractants and repellents. *Ann. Rev. Entomol.* **11**, 403.

Jacobson, M. (1969). Sex pheromone of the pink bollworm moth: biological masking by its geometrical isomer. *Science* **163**, 190.

Jacobson, M., and Beroza, M. (1963). Chemical insect attractants. *Science* **140**, 1367.

Jacobson, M., and Beroza, M. (1964). Insect attractants. *Sci. Am.* **211**, (No. 2) 20.

Jacobson, M., and Harding, C. (1968). Insect sex attractants. IX. Chemical conversion of the sex attractant of the cabbage looper to the sex pheromone of the fall armyworm. *J. Econ. Entomol.* **61**, 394.

Jacobson, M., Toba, H. H., Debolt, J., and Kishaba, A. N. (1968). Insect sex attractants. VIII. Structure-activity relationships in sex attractant for male cabbage loopers. *J. Econ. Entomol.* **61**, 84.

Jefferson, R. N., Shorey, H. H., and Rubin, R. E. (1968). Sex pheromones of noctuid moths. XVI. The morphology of the female sex pheromone glands of eight species. *Ann. Entomol. Soc. Am.* **61**, 861.

Jones, W. A., and Jacobson, M. (1968). Isolation of *N,N*-diethyl-*m*-toluamide (Deet) from female pink bollworm moths. *Science* **159**, 99.

Jones, W. A., Jacobson, M., and Martin, D. F. (1966). Sex attractant of the pink bollworm moth: isolation, identification and synthesis. *Science* **152**, 1516.

Karlson, P., and Butenandt, A. (1959). Pheromones (ectohormones) in insects. *Ann. Rev. Entomol.* **4**, 39.

Kaufmann, T. (1968). Observations on the biology and behavior of the evergreen bagworm moth, *Thyridopteryx ephemeraeformis* (Lepidoptera: Psychidae). *Ann. Entomol. Soc. Am.* **61**, 38.

Keys, R. E., and Mills, R. B. (1968). Demonstration and extraction of a sex attractant from female Angoumois grain moths. *J. Econ. Entomol.* **61**, 46.

Lindquist, O. H., and Bowser, R. L. (1966). A biological study of the leaf miner, *Chrysopeleia ostryaella* Chambers (Lepidoptera: Cosmopterygidae), on ironwood in Ontario. *Can. Entomologist* **98**, 252.

McCabe, E. T., Barthel, W. F., Gertler, S. I., and Hall, S. A. (1954). Insect repellents. III. *N,N*-diethylamides. *J. Org. Chem.* **19**, 493.

McDonough, L. M., George, D. A., and Butt, B. A. (1969a). Artificial sex pheromones for the codling moth. *J. Econ. Entomol.* **62**, 243.

McDonough, L. M., George, D. A., Butt, B. A., Jacobson, M., and Johnson, G. R. (1969b). Isolation of a sex pheromone of the codling moth. *J. Econ. Entomol.* **62**, 62.

Meinwald, J., and Meinwald, Y. C. (1966). Structure and synthesis of the major components in the hairpencil secretion of a male butterfly, *Lycorea ceres ceres* (Cramer). *J. Am. Chem. Soc.* **88**, 1305.

Meinwald, J., Meinwald, Y. C., Wheeler, J. W., Eisner, T., and Brower, L. P. (1966). Major components in the exocrine secretion of a male butterfly *(Lycorea)*. *Science* **151**, 583.

Pattenden, G. (1968). A synthesis of propylure, sex attractant of the pink bollworm moth.. *J. Chem. Soc.* C 2385.

Rahn, R. (1966). La teigne du poireau *Acrolepia assectella* Zeller, éléments de biologie et mise au point d'avertissements agricoles fondes sur le piegeage sexuel des males. *Proc. Acad. Agr. France* p.997.

Read, J. S., Warren, F. L., and Hewitt, P. H. (1968). Identification of the sex pheromone of the false codling moth. *Chem. Commun.* p.792.

Regnier, F. E., and Law, J. H. (1968). Insect pheromones. *J. Lipid Res.* **9**, 541.

Riddiford, L. M. (1967). *trans*-2-Hexenal: mating stimulant for polyphemus moths. *Science* **158**, 139.

Riddiford, L. M., and Williams, C. M. (1967a). Volatile principle from oak leaves: role in sex life of the polyphemus moth. *Science* **155**, 589.

Riddiford, L. M., and Williams, C. M. (1967b). Chemical signalling between polyphemus moths and between moths and host plant. *Science* **156**, 541.

Roelofs, W. L. (1966). Sex attractants used to combat insects. *Farm Res. (N. Y.)* **32**, (No. 2) 2.

Roelofs, W. L., and Arn, H. (1968a). Red-banded leaf roller sex attractant characterized. *Food Life Sci.* **1**, (No. 1) 12.

Roelofs, W. L., and Arn, H. (1968b). Sex attractant of the red-banded leaf roller moth. *Nature* **219**, 513.

Roelofs, W. L., and Comeau, A. (1968). Sex pheromone perception. *Nature* **220**, 600.

Roelofs, W. L., and Feng, K. C. (1967). Isolation and bioassay of the sex pheromone of the red-banded leaf roller, *Argyrotaenia velutinana* (Lepidoptera: Tortricidae). *Ann. Entomol. Soc. Am.* **60**, 1199.

Roelofs, W. L. and Feng, K. C. (1968). Sex pheromone specificity tests in the Tortricidae—an introductory report. *Ann. Entomol. Soc. Am.* **61**, 312.

Roth, L. (1967). Male pheromones. *In* "McGraw-Hill Yearbook of Science and Technology," pp. 193-194. McGraw-Hill, New York.

Saringer, G., Wegh, G., and Rada, K. (1968). Sexual attractiveness of virgin plum fruit moth, *Grapholitha funebrana* Tr. (Lepidoptera: Tortricidae) females examined by ^{32}P labelled males. *Acta Phytopathol. Acad. Sci. Hung.* **3**, 373.

Schneider, D. (1964). Insect antennae. *Ann. Rev. Entomol.* **9**, 103.

Schneider, D. (1969). Insect olfaction: Deciphering system for chemical messages. *Science* **163**, 1031.

Sekul, A. A., and Cox, H. C. (1965). Sex pheromone in the fall armyworm, *Spodoptera frugiperda* (J. E. Smith). *Bioscience* **15**, 670.

Sekul, A. A., and Cox, H. C. (1967). Response of males to the female sex pheromone of the armyworm, *Spodoptera frugiperda* (Lepidoptera: Noctuidae): a laboratory evaluation. *Ann. Entomol. Soc. Am.* **60**, 691.

Sekul, A. A., and Sparks, A. N. (1967). Sex pheromone of the fall armyworm moth: isolation, identification, and synthesis. *J. Econ. Entomol.* **60**, 1270.

Shorey, H. H., and Gaston, L. K. (1967). Pheromones. *In* "Pest control: biological, physical, and selected chemical methods" (W. W. Kilgore and R. L. Doutt, eds.), pp. 241-265. Academic Press, New York.

Shorey, H. H., Gaston, L. K., and Fukuto, T. R. (1964). Sex pheromones of noctuid moths, I: A quantitative bioassay for the sex pheromone of *Trichoplusia ni* (Lepidoptera: Noctuidae). *J. Econ. Entomol.* **57**, 252.

Shorey, H. H., Gaston, L. K., and Roberts, J. S. (1965). Sex pheromones of noctuid moths. VI. Absence of behavioral specificity for the female sex pheromones of *Trichoplusia ni* versus *Autographa californica,* and *Heliothis zea* versus *H. virescens* (Lepidoptera: Noctuidae). *Ann. Entomol. Soc. Am.* **58**, 600.

Shorey, H. H., Gaston, L. K., and Saario, C. A. (1967). Sex pheromones of noctuid moths. XIV. Feasibility of behavioral control by disrupting pheromone communication in cabbage loopers. *J. Econ. Entomol.* **60**, 1541.

Shorey, H. H., McFarland, S. U., and Gaston, L. K. (1968). Sex pheromones of noctuid moths. XIII. Changes in pheromone quantity, as related to reproductive age and mating history, in females of seven species of Noctuidae (Lepidoptera). *Ann. Entomol. Soc. Am.* **61**, 372.

Solomon, J. D., and Morris, R. C. (1966). Sex attraction of the carpenterworm moth. *J. Econ. Entomol.* **59**, 1534.

Toba, H. H., Kishaba, A. N., and Wolf, W. W. (1968). Bioassay of the synthetic female sex pheromone of the cabbage looper. *J. Econ. Entomol.* **61**, 812.

Traynier, R. M. M. (1968). Sex attraction in the Mediterranean flour moth, *Anagasta kühniella*: location of the female by the male. *Can. Entomologist* **100**, 5.

Tunstall, J. P. (1965). Sex attractant studies in *Diparopsis. Pest Articles News Summaries 11A*, 212.

Warthen, D. (1968). Synthesis of *cis*-9-tetradecen-l-ol acetate, the sex pheromone of the fall armyworm. *J. Med. Chem.* **11**, 371.

Warthen, D., and Jacobson, M. (1967). Insect sex attractants. VII. 5,9-Tridecadien-l-ol acetates. *J. Med. Chem.* **10**, 1190.

Warthen, D., and Jacobson, M. (1968). Insect sex attractants. X. 5-Dodecen-l-ol acetates, analogs of the cabbage looper sex attractant. *J. Med. Chem.* **11**, 373.

Weatherston, J., and Percy, J. E. (1968). Studies of physiologically active arthropod secretions. I. Evidence for a sex pheromone in female *Vitula edmandsae* (Lepidoptera: Phycitidae). *Can. Entomologist* **100**, 1065.

Wilson, E. O. (1963). Pheromones. *Sci. Am.* **208**, (No. 5) 100.

ATTRACTANT PHEROMONES OF COLEOPTERA

Robert M. Silverstein

I. INTRODUCTION

Coleoptera, the largest of all orders of insects, consists of the beetles and the weevils. Their total economic impact on man is almost incalculable. They damage and destroy field crops and fruit, shade, and forest trees as well as food and fibers in storage and in homes. However, beneficial insects (e.g., the familiar lady beetles) also occur in this order.

The phenomenon of pheromone communication has been reported for many beetles and weevils, but only in a few cases have the pheromones been isolated and chemically identified. All these compounds may be classified as attractants, i.e., they induce movement of the insect toward the source of the chemical. Since the ultimate purpose of such movement of one or both sexes would seem to be propagation of the species, the designation "sex attractant" has been applied—and this designation has been vigorously attacked and defended. Indeed, research in this field seems marked by lively and occasionally intemperate polemics [for an example of a "Coleopteran" exchange see Vite (1967) vs. Wood *et al.* (1967a)]. Reasons for such exchanges lie in the difficulties involved in isolating and identifying minute amounts of biologically active compounds from complex substrates, in the complications caused by synergistic and masking effects, and in the problems of interpreting insect behavior, and of course in the personalities of the investigators.

Attractant pheromones have been identified in eight insects in four Families of the order Coleoptera.

Scolytidae

> *Ips confusus* (LeConte) (California five-spined *ips*)
> *Dendroctonus brevicomis* LeConte (western pine beetle)
> *Dendroctonus ponderosae* Hopkins (mountain pine beetle)
> *Dendroctonus frontalis* Zimmerman (southern pine beetle)

Dermestidae

> *Attagenus megatoma*(Fabricius) (black carpet beetle)
> *Trogoderma inclusum* LeConte

Elateridae

> *Limonius californicus* (Mannerheim) (sugar beet wireworm)

Curculionidae

> *Anthonomus grandis* Boheman (Cotton boll weevil)

The attractant pheromones identified thus far in the order Lepidoptera have been long-chain esters or alcohols with one or two double bonds, and in one case a branched chain (see chapter by Jacobson *et al.*) The attractant pheromones in the order Coleoptera are more varied in structure.

This chapter describes briefly the isolation, identification, and synthesis of the chemical compounds comprising the attractant pheromones of the coleopteran insects that have been reported to date. The review is critical rather than encyclopedic. Controversial issues are noted and an attempt is made to evaluate some of the claims. The author's expressed opinions on chemical issues are his own, and he would prefer to stop at that point. Since the chemical and biological problems are inseparable, however, he has sought counsel from his collaborators, Dr. David L. Wood (Department of Entomology and Parasitology, University of California, Berkeley), Dr. Wendell E. Burkholder, Department of Entomology, University of Wisconsin, and Dr. William D. Bedard (Pacific Southwest Forest and Range Experiment Station, U. S. Forest Service, Berkeley, California). Nonetheless, all statements in this chapter remain the responsibility of the author whose position as a participant in some of the controversies should be noted.

II. TERPENE ALCOHOLS COMPRISING THE ATTRACTANT PHEROMONE OF *Ips confusus*

A. HISTORY AND BACKGROUND

The first report of the isolation and identification of an attractant pheromone of a coleopteran insect was presented at the 16th Annual Meeting of the Entomological Society of Canada at Banff, September 1966 (Silverstein *et al.,* 1966a). Results of the laboratory bioassay and field tests on natural populations were also presented and are recorded in the abstracts of the meeting. The attractant in the frass of the male beetle boring in ponderosa pine attracts both males and females (Wood, 1962; Wood *et al.,* 1966). Three terpene alcohols were identified as the principal components of the attractant (Silverstein *et al.,* 1966b): (−)-2-methyl-6-methylene-7-octen-4-ol (I), (+)-*cis*-verbenol (II), and (+)-2-methyl-6-methylene-2,7-octadien-4-ol (III). From 4.5 kg of frass, about 250 mg of (I), 1.5 mg of (II), and 50 mg of (III) were isolated in pure form.

(I)   (II)   (III)

None of these compounds was attractive by itself, but in the laboratory bioassay, a mixture of compounds (I) and (II) or (I) and (III) evoked a strong walking response which was reinforced by addition of the third component (Wood *et al.,* 1967b). In the field, natural populations of flying beetles responded weakly to a mixture of (I) and (III), and strongly to a mixture of the three compounds (Wood *et al.,* 1968). Another species *Ips latidens* was attracted to compound (I) and a mixture of (I) and (II) in the laboratory and in the field, and its response was eliminated by addition of compound (III) in the laboratory (Wood *et al.,* 1967b). These unique synergistic and blocking effects were striking. In further field tests with the three synthesized compounds, two predators of bark beetles, *Enoclerus lecontei* (Wolcott) (Coleoptera: Cleridae) and *Temnochila virescens chlorodia* (Mannerheim) (Coleoptera: Ostomidae), also responded (Wood *et al.,* 1968). The implication is that these predators may use the chemical messengers produced by the bark beetles to find high prey densities. In these field studies 1.5 mg of (I), 1 mg of (II), and 0.5 mg of (III) were delivered together over a period of about 6 hours. Logging slash one kilometer downwind was thought to be the source of the beetle population.

Cross attraction among other species of *Ips* to each other's frass has been found in the laboratory and in the field (Wood *et al.,* 1968). Probably most of this communication depends on the same, or closely related, terpene alcohols.

Pitman *et al.* (1966) reported the isolation from "hindguts" (presumably the fecal contents since the beetles had been removed from feeding tunnels) of a major gas chromatographic fraction, which elicited a response from walking beetles. He concluded that "the data do not contradict the conclusion that the pheromone is a single compound." This "compound" is described by Renwick (1966) as a tertiary terpene alcohol. Apparently this "compound" was a gross mixture containing at least two of the active compounds.

In a subsequent letter to *Science,* Vite (1967) (1) denied the validity of using walking beetles in a laboratory bioassay to follow the steps in a chemical isolation; (2) objected to the term sex attractant; (3) advocated "pure fecal pellets or dissected hindguts" rather than frass as a source of attractants; and (4) postulated that "only experiments performed under field conditions can prove whether the conclusions presented by Silverstein *et al.* are actually valid and do not fall prey to deficiencies in the laboratory bioassay." In reply, Wood *et al.* (1967a) (1) agreed that "walking responses in the laboratory may not precisely reflect the behavior of flying populations," insisted, however, that "a laboratory bioassay is an indispensable tool for following the isolation of minute amounts of active components" provided that such results are correlated with field data, and cited use of the laboratory bioassay in Vite's own studies; (2) acknowledged the unsatisfactory state of the terminology used to describe the attraction phenomenon; (3) defended their preference for frass over hindguts on the grounds that beetles are attracted to frass in the field and have not been observed to fly or crawl toward, or into, the hindgut of another intact beetle of either sex; and (4) noted that Vite was present at the Banff meeting at which results of successful field experiments were presented. Two additional comments may also be made: (1) claims by Vite's group (Pitman and Vite, 1963) to the discovery of secretory areas in the hindgut had to be retracted (Pitman *et al.,* 1965); and (2) claims by Vite's group to "independent identification" of the attractants from hindguts of *Ips confusus* should be retracted since not only were the claims false, but they were published (*Sci. Am.,* 1966) 3 months after the Banff meeting.

B. GENERAL METHODOLOGY OF ISOLATION AND IDENTIFICATION

The general methodology involved in isolating and identifying insect pheromones has been described with particular reference to studies of the attractants produced by *Ips confusus* (Silverstein *et al.,* 1967a). The following steps are considered essential.

1. Development of a sensitive bioassay that requires only a small amount of attractant. Usually this means a laboratory bioassay, which should be based on sound biological principles derived from the behavior of natural populations.

2. Production of large amounts of starting material either by mass rearing or by large-scale field collecting.

3. Isolation of each compound in a state of high purity. Each step of the isolation must be monitored by the bioassay.

4. Identification of the individual compounds. Since these are usually obtained in milligram or microgram quantities, identification rests heavily on spectrometric evidence (Silverstein and Bassler, 1967).

5. Confirmation of postulated structures by comparison with rationally synthesized compounds.

6. Evaluation of biological activity of the synthesized compounds in the laboratory and under natural field conditions.

The complications of an isolation procedure are demonstrated (Silverstein *et al.*, 1967a) by the disappearance of attractant activity as the fractionation steps became more refined, and its reappearance on recombination of several of the fractions. These synergistic effects emphasize the importance of monitoring each step by bioassay; no fraction can be discarded until it has been tested in combination with the other fractions. Some idea of the tediousness of the isolation procedure may be gained by noting that six sequential fractionations by gas chromatography were needed to obtain a pure sample of one of the terpene alcohols (compound II). Gas chromatography had been preceded by high-vacuum distillation and silica gel chromatography of an extract of frass.

Identification of the three terpene alcohols was achieved [in the case of compound (II) on a 1.5 mg sample] by correlating the complementary information obtained from four spectra: mass, infrared, nuclear magnetic resonance, and ultraviolet. This was possible even though compounds (I) and (III) were novel compounds, i.e., not previously recorded in the chemical literature. *cis*-Verbenol has been recorded, but no spectral data had been reported, and even its melting point was in dispute.

C. SYNTHESIS OF THE Ips confusus COMPOUNDS

The attractant terpene alcohols were synthesized by the following rational sequences (Reece *et al.*, 1968).

(I)

Verbenone

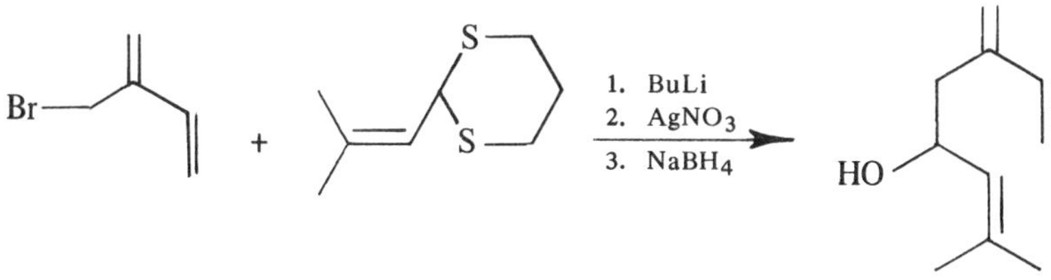

(III)

III. ATTRACTANTS OF *Dendroctonus frontalis*, *Dendroctonus brevicomis*, AND *Dendroctonus ponderosae*

A. *trans-VERBENOL AND VERBENONE*

The history of the isolation and identification of *trans*-verbenol and verbenone from the hindgut (contents) of several different bark bettles demonstrates the confusion caused by failure to isolate components associated with the type of biological activity sought. *trans*-Verbenol and verbenone were isolated and identified from the hindguts of the female and male, respectively, of *D. frontalis* and *D. brevicomis* (Renwick, 1967). The apparent rationale for this approach is to search for sex-specific compounds. A small amount of *trans*-verbenol, however, was later shown to be present in *D. frontalis* and *D. brevicomis* males, and verbenone in the female of these two species (Pitman *et al.*, 1969). The biological significance of these observations is not clear. Verbenone "appears innocuous" (Pitman *et al.*, 1968). On the basis of two series of three 10-minute field tests against *D. frontalis,* Kinzer *et al.* (1969) report that "the response to the synthetic attractant [frontalin] was increased by adding oleoresin. . . and synthetic *trans*-verbenol. . .". In one series, a mixture of frontalin, *trans*-verbenol, and oleoresin was more attractive (average catch, 10.7 beetles) than frontalin (1) or oleoresin (0) or crushed male and female beetles plus oleoresin (4.3). In the second series, a mixture of frontalin and oleoresin was about as attractive (10.7) as crushed beetles and oleoresin (11.3) and more attractive than a mixture of oleoresin and *trans*-verbenol (0). No test results were shown for a mixture of frontalin and *trans*-verbenol, nor was the mixture of frontalin, *trans*-verbenol, and oleoresin compared with the mixture of frontalin and oleoresin in either series. Thus, no direct assessment of the role of *trans*-verbenol can be made from these tests.

trans-Verbenol sprayed on trees under attack by *D. frontalis* increased the landing rate of *D. frontalis* (Vite and Crozier, 1968), and, when added to crushed *D. brevicomis* males, it seemed to increase the catch of *D. frontalis* over that of crushed males alone (Renwick and Vite, 1968). The latter data were republished by Pitman *et al.* (1969) with the addition of the qualifying adverb "seemingly."

trans-Verbenol was also found in the female *D. ponderosae*, and when sprayed on a "marginally" attractive billet (containing introduced females), it enhanced the attractiveness of the billet for *D. ponderosae*, although when sprayed on an uninfested billet no attraction resulted (Pitman *et al.*, 1968). Subsequently it was found that a mixture of *trans*-verbenol and oleoresin was attractive to both sexes (Pitman and Vite, 1969).

OH   O
(IV)   (V)

The field studies by Pitman and co-workers present very little description of the methodology employed and seem to imply that all field experiments utilize naturally occurring populations. In a workshop presentation, however, Pitman (1969) described the source of the beetles for a series of field tests as a truckload of beetle-infested bark

(*D. brevicomis*) parked downwind from the trap. When the bark warms up and the swarms of beetles emerge, the traps are baited. A single test may last only several minutes. Since these field studies do not challenge the unmanipulated natural population, the results may not give an adequate estimate of the response to potential pheromone compounds.

Both *trans*-verbenol (IV) and verbenone (V) are well-known compounds. Verbenone is obtained from natural sources, and *trans*-verbenol was synthesized by oxidation of α-pinene.

B. BREVICOMIN

1. History and Isolation

The attractant in frass produced by newly emerged female *Dendroctonus brevicomis* boring in ponderosa pine attracts both males and females. From 1.6 kg of frass, about 1.5 mg of a rather volatile compound was isolated that was strongly active by itself in the laboratory bioassay and was synergized by a hydrocarbon fraction that was inactive by itself. The active compound was identified as a unique bicyclic ketal structure: *exo*-7-ethyl-5-methyl-6,8-dioxabicyclo[3.2.1]octane (VI) (Silverstein *et al.,* 1968).

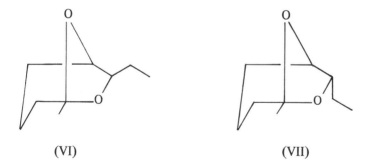

(VI)   (VII)

This compound was assigned the trivial name brevicomin, which was later modified to *exo*-brevicomin, because the epimeric structure, named *endo*-brevicomin (VII) was also found to be present in frass. The [3.2.1]bicyclic ketal ring system is a novel structure in natural products, and indeed the first report of an analogous compound (in hop oil) (Naya and Kotake, 1967) was called to the attention of the above authors only recently.

One of the synergistic components of the hydrocarbon fraction (from the host tree) was isolated and identified as myrcene. In the laboratory, *exo*-brevicomin was highly attractive to male *D. brevicomis,* but the addition of myrcene further increased the response. Whereas *endo*-brevicomin was only marginally attractive in the laboratory to both sexes, a mixture of it and myrcene was highly attractive (Wood, unpublished data, 1969). Field studies on natural populations, using 2 mg or 20 mg of *exo*-brevicomin mixed with 20 mg of myrcene delivered over a 6-hour period, confirmed the synergism noted in the laboratory bioassay (Bedard *et al.,* 1969a). *exo*-Brevicomin was moderately attractive to both sexes of *D. brevicomis* and highly attractive to both sexes of the bark beetle predator, *Temnochila virescens chlorodia* (Bedard *et al.,* 1969). Vite and Pitman (1969 noted the predator response, but were unable to show a response by *D. brevicomis* to synthetic brevicomin. When it was added to ponderosa pine oleoresin (of which myrcene is a component), however, a strong response was evoked.

2. Identification and Synthesis of exo- and endo-Brevicomin

A bicyclic ketal structure was postulated from the mass, infrared, nuclear magnetic resonance, and ultraviolet spectra. The questions of ring size and substituents were resolved by the results of a hydrogenolysis experiment on a 50 μg sample in a modified Beroza "carbon skeleton determinator" (Beroza and Acree, 1964; Brownlee and Silverstein, 1968). The major hydrogenolysis product was separated by gas chromatography and identified as *n*-nonane by mass spectrometry. The geometry of the ethyl substituent was elucidated from the nuclear magnetic resonance spectra, since, in the exo form, two vicinal protons are almost at right angles to each other and thus show virtually zero coupling.

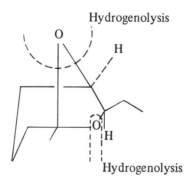

exo-Brevicomin (VI)

Confirmation of the postulated structures for both *exo*-brevicomin and *endo*-brevicomin followed from the stereospecific synthetic sequence (Bellas *et al.,* 1969) (Fig. 1).

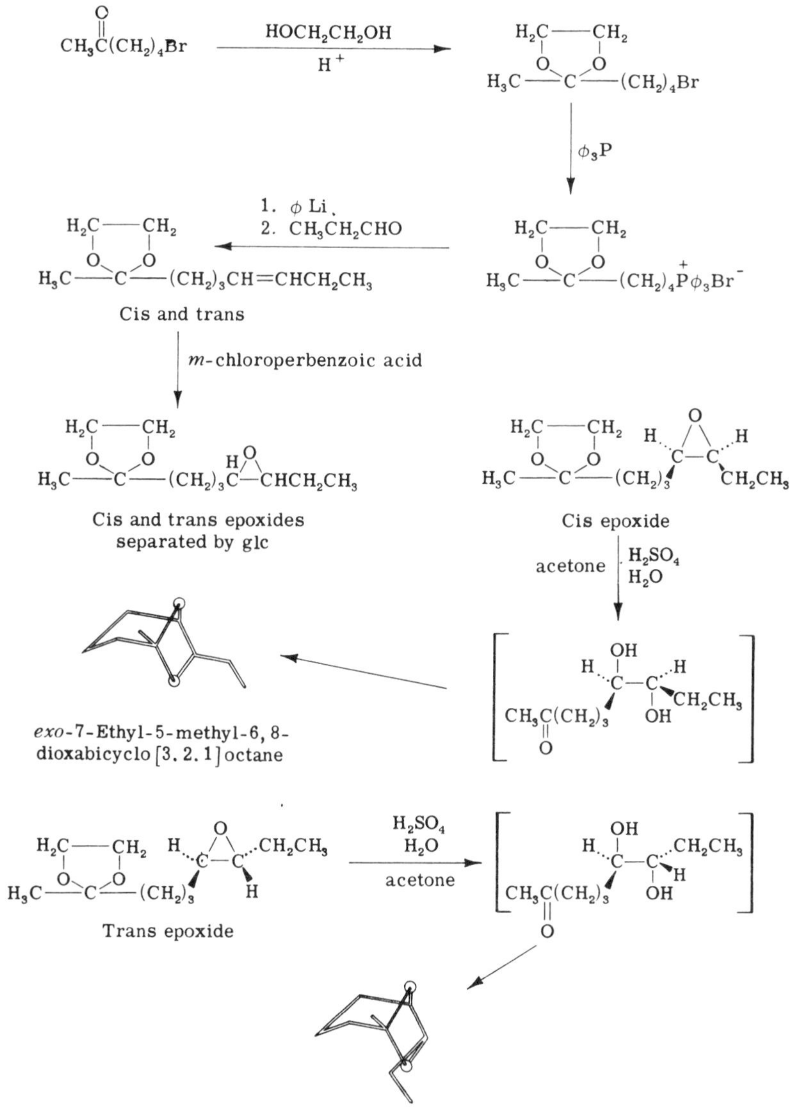

exo-7-Ethyl-5-methyl-6, 8-dioxabicyclo[3.2.1]octane

$endo$-7-Ethyl-5-methyl-6, 8-dioxabicyclo[3.2.1]octane

FIG. 1. Synthesis of *exo*- and *endo*-brevicomin.

Wasserman and Barber (1969) in a study of [3.2.1]ring systems independently synthesized *exo*-brevicomin (VI) by epoxidation of *cis*-non-6-en-2-one followed by cyclization. This is currently the synthesis of choice. Spectra and biological activity of the material synthesized by both procedures are in complete agreement.

C. FRONTALIN

1. History and Isolation

Kinzer *et al.* (1969) isolated 1,5-dimethyl-6,8-dioxabicyclo[3.2.1]octane (VIII) as a prominent component from the hindgut contents of the male *D. brevicomis* while examining the gas chromatographic region in which the previously described brevicomin had been shown to occur (Silverstein *et al.*, 1968). No biological significance relative to *D. brevicomis* was ascribed to this compound by Kinzer *et al.*, and no acknowledgment was made of the close chemical relationship between this bicyclic ketal and brevicomin, and the biogenic and taxonomic implications thereof. Vite (1968) states, however, that this new analog of brevicomin "proved to be highly attractive to the western pine beetle [*D. brevicomis*]."

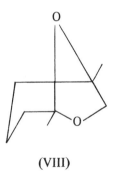

(VIII)

A minor fraction isolated from hindguts of *D. frontalis* females had previously been shown to be attractive to both sexes in both the laboratory and field (Renwick and Vite, 1968). The investigators were unable to identify the active component, but it was later established that the active component in female *D. frontalis* and the bicyclic ketal (VIII) isolated from male *D. brevicomis* had the same gas chromatographic properties (Kinzer *et al.*, 1969; Pitman *et al.*, 1969). The synthesized compound (VIII) is reported to attract small numbers of *D.*

frontalis in the field, but is more attractive when presented with oleoresin (Kinzer *et al.,* 1969). However, the compound was inactive in the laboratory (Vite, 1968)—this result contrasting with the reported activity of the material isolated from hindguts of *D. frontalis* females (Renwick and Vite, 1968). These results notwithstanding, compound (VIII) was named frontalin, contrary to the usual practice of assigning a trivial name descriptive of the source (i.e., *D. brevicomis*) from which the compound was identified and in which it shows biological activity (Vite, 1968; Bedard *et al.,* 1969b). In view of this confusing history and in the absence of a rational, sequential, monitored isolation procedure, it is difficult, at this time, to accept the designation of frontalin as the "principal component of the aggregating pheromone" of *D. frontalis*.

2. Identification and Synthesis of Frontalin

Comparison of the spectral data for frontalin with reference spectra for brevicomin allowed ready recognition of the bicyclic ketal system. There was no problem of configuration to be resolved.

Frontalin was synthesized as follows.

IV. ATTRACTANTS OF *Attagenus megatoma*

A. HISTORY

The response of the male *Attagenus megatoma* beetle to the unmated female has been described by Burkholder and Dicke (1966), whose bioassay was used to monitor the isolation steps.

The sex attractant is *trans*-3, *cis*-5-tetradecadienoic acid (IX), named megatomoic acid, 4 mg which was isolated from 8000 virgin females (Silverstein *et al.,* 1967b). No other fraction in the various stages of isolation elicited a response, nor were synergistic or masking effects noted.

$$\text{CH}_3(\text{CH}_2)_7 \overset{\text{H}}{\text{C}}=\overset{\text{H}}{\text{C}}-\text{C}=\overset{\text{H}}{\text{C}}-\text{CH}_2\text{COOH}$$
$$\underset{\text{H}}{}$$

(IX)

Field studies, carried out in a large room, showed that milligram quantities of megatomoic acid could attract male beetles over a distance of about 25 feet. Experiments are in progress to test the possibilities of "confusing" the beetles with a relatively high concentration of megatomoic acid with the aim of disrupting mating activities.

B. ISOLATION, IDENTIFICATION, AND SYNTHESIS OF MEGATOMOIC ACID

The sequence of isolation consisted of extraction of female beetles with benzene in a Waring Blendor, removal of solvent, short-path, high-vacuum distillation of the residue, partition into alkaline solution, regeneration and extraction of the free acid, silica gel chromatography, anion-exchange chromatography, esterification with diazomethane, and gas chromatography of the methyl ester. Bioassays were carried out on saponified aliquots of the gas chromatographic fractions.

The compound was synthesized as follows (Silverstein *et al.*, 1967b).

$$\text{CH}_3(\text{CH}_2)_7\text{C}\!\equiv\!\text{CH} \xrightarrow[\text{2. CH}_2\text{=CHCHO}]{\text{1. EtMgBr}} \text{CH}_2(\text{CH}_2)_7\text{C}\!\equiv\!\text{C}-\overset{\overset{\text{OH}}{|}}{\text{CH}}-\text{CH=CH}_2$$

$$\xrightarrow{\text{PBr}_3} \text{CH}_3(\text{CH}_2)_7\text{C}\!\equiv\!\text{C}-\text{CH=CH}-\text{CH}_2\text{Br} \xrightarrow[\substack{\text{2. MeOH/HCl} \\ \text{3. OH}^-}]{\text{1. CuCN}} \text{(IX)}$$

The other geometric isomers were synthesized and were found to be very much less active, if at all; the small amount of activity may have been due to trace contamination by the active isomer.

V. ATTRACTANT COMPOUNDS OF *Trogoderma inclusum*

A. HISTORY

The response of the *Trogoderma inclusum* male beetle to the female has been demonstrated by Burkholder and Dicke (1966), whose laboratory bioassay was used to monitor the isolation of the attractant compounds.

Two attractant compounds were isolated from the female beetles (Rodin *et al.*, 1969): (−)-14-methyl-*cis*-8-hexadecen-1-ol (X) and (−)-methyl 14-methyl-*cis*-8-hexadecenoate (XI).

$$CH_3CH_2\overset{\overset{\displaystyle CH_3}{|}}{C}H(CH_2)_4\overset{\overset{\displaystyle H}{}}{C}=\overset{\overset{\displaystyle H}{}}{C}(CH_2)_6CH_2OH$$

(X)

$$CH_3CH_2\overset{\overset{\displaystyle CH_3}{|}}{C}H(CH_2)_4\overset{\overset{\displaystyle H}{}}{C}=\overset{\overset{\displaystyle H}{}}{C}(CH_2)_6COOCH_3$$

(XI)

These compounds show the anteiso chain branching which may be derived biogenetically from α-methylbutyric acid.

Each compound (synthesized) is active by itself, the alcohol more so than the ester. During the isolation, at least two other fractions showed some attractant activity. The attractant compounds do not synergize one another; their effect seems additive.

Interspecies responses for five species of male *Trogoderma* to extracts of females of seven species have been shown (Vick *et al.*, 1969). Similar interspecies response has been shown toward each of the synthesized attractants.

B. ISOLATION, IDENTIFICATION, AND SYNTHESIS

Unmated female beetles were extracted with benzene in a Waring Blendor. The soluble material (after solvent removal) was distilled in a short-path apparatus under high vacuum. Acidic components were removed by alkaline extraction, and the neutral material was fractionated by silica gel chromatography followed by gas chromatography. About 3 mg of (X) and 1 mg of (XI) were isolated from 100,000 beetles.

Spectral evidence indicated that (X) was a C_{17} primary alcohol with one double bond (cis). Ozonolysis (Beroza and Bierl, 1966) gave an aldehyde that eluted from a gas chromatographic column between the reference compounds *n*-octanal and *n*-nonanal. Hydrogenolysis gave two products, which were separated by gas chromatography and identified as 3-methylpentadecane and 3-methylhexadecane.

The spectral evidence indicated that (XI) was a C_{17} methyl ester with one double bond (cis). Ozonolysis yielded the same aldehyde that was obtained on ozonolysis of (X); thus, the positions of the double bond and the branch were fixed.

Compounds (X) and (XI) were synthesized as follows (Rodin *et al.,* 1969).

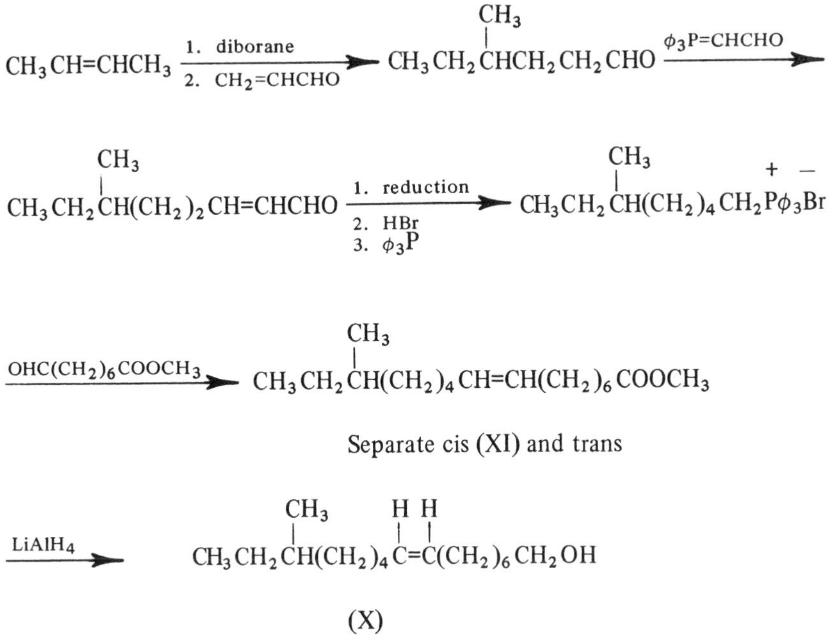

$$CH_3CH=CHCH_3 \xrightarrow[\text{2. } CH_2=CHCHO]{\text{1. diborane}} CH_3CH_2\overset{\overset{\displaystyle CH_3}{|}}{C}HCH_2CH_2CHO \xrightarrow{\phi_3P=CHCHO}$$

$$CH_3CH_2\overset{\overset{\displaystyle CH_3}{|}}{C}H(CH_2)_2CH=CHCHO \xrightarrow[\substack{\text{2. HBr} \\ \text{3. } \phi_3P}]{\text{1. reduction}} CH_3CH_2\overset{\overset{\displaystyle CH_3}{|}}{C}H(CH_2)_4CH_2\overset{+}{P}\phi_3\overset{-}{Br}$$

$$\xrightarrow{OHC(CH_2)_6COOCH_3} CH_3CH_2\overset{\overset{\displaystyle CH_3}{|}}{C}H(CH_2)_4CH=CH(CH_2)_6COOCH_3$$

Separate cis (XI) and trans

$$\xrightarrow{LiAlH_4} CH_3CH_2\overset{\overset{\displaystyle CH_3}{|}}{C}H(CH_2)_4\overset{\overset{\displaystyle H}{|}}{C}=\overset{\overset{\displaystyle H}{|}}{C}(CH_2)_6CH_2OH$$

(X)

VI. ATTRACTANT OF *Limonius californicus*

An attractant produced in large amounts (>100 μg per insect) by the female sugar beet wireworm (*Limonius californicus*), which attracts the male, has been identified as valeric acid (Jacobson *et al.,* 1968). Apparently the attractant is stored in bound form and is released on demand in small amounts. Activity was demonstrated in laboratory bioassays and in the field. A number of homologous acids and a variety of other chemical compounds showed no attraction. Caproic acid was slightly attractive.

ISOLATION AND IDENTIFICATION

An ether extract of 18 female abdomens was chromatographed on silica acid, and the active fraction was extracted into alkali. Liquid and solid acids were separated and the liquid acid fraction was distilled. The distillate gave a single spot on paper chromatography that was coincident with that of valeric acid. Spectral data on a gas chromatographic fraction of the methyl ester confirmed the identity of the attractant. Each step of the isolation procedure was monitored with a laboratory bioassay.

VII. TERPENOID ATTRACTANTS OF *Anthonomus grandis* *(COTTON BOLL WEEVIL)*

Four terpenoids comprise the attractant secreted by the male *Anthonomus grandis* Boheman, the cotton boll weevil. This study is described in the chapter by J. H. Tumlinson *et al.*

SUMMARY

The isolation and identification of pheromones from Coleoptera and practical considerations concerned with their bioassay and use are described.

Specifically, pheromones from *Ips confusus, Dendoctonus frontalis, D. brevicomis, D. ponderosae, Attagenus megatoma, Trogoderma inclusum,* and *Limonius californicus* are discussed.

REFERENCES

Bedard, W. D., Tilden, P. E., Wood, D. L., Silverstein, R. M., Brownlee, R. G., and Rodin, J. O. (1969a). Western pine beetle: Field response to its sex pheromone and a synergistic host terpene, myrcene. *Science* **164**, 1284.

Bedard, W. D., *et al* (1969b). In preparation.

Bellas, T. E., Brownlee, R. G., and Silverstein, R. M. (1969). Synthesis of brevicomin, principal sex attractant in the frass of the female western pine beetle. *Tetrahedron* **25**, 5149.

Beroza, M., and Acree, F., Jr. (1964). A new technique for determining chemical structure by gas chromatography. *J. Assoc. Offic. Agr. Chem.* **47**, 1.

Beroza, M., and Bierl, B. A. (1966). Apparatus for ozonolysis of microgram to milligram amounts of compound. *Anal. Chem.* **38**, 1976.

Brownlee, R. G., and Silverstein, R. M. (1968). A micro-preparative gas chromatograph and a modified carbon skeleton determinator. *Anal. Chem.* **40**, 2077.

Burkholder, W. E., and Dicke, R. J. (1966). Evidence of sex pheromones in females of several species of Dermestidae. *J. Econ. Entomol.* **59**, 540.

Jacobson, M., Lilly, C. E., and Harding, C. (1968). Sex attractant of the sugar beet wireworm: identification and biological activity. *Science* **159**, 208.

Kinzer, G. W., Fentiman, A. F., Jr., Page, T. E., Jr., Foltz, R. L., Vite, J. P., and Pitman, G. B. (1969). Bark beetle attractants: identification, synthesis, and field bioassay of a new compound isolated from *Dendroctonus. Nature* **221**, 477.

Naya, Y., and Kotake, M. (1967). A new constituent of hop oil. *Tetrahedron Letters* p.2459.

Pitman, G. B. (1969). Forest Entomology Seminar, Forest and Range Station, U. S. Forest Service, Berkeley, California.

Pitman, G. B., and Vite, J. P. (1963). The pheromone of *Ips confusus. Contrib. Boyce Thompson Inst.* **22**, 221.

Pitman, G. B., and Vite, J. P. (1969). Aggregation behavior of *Dendroctonus ponderosae* (Coleoptera: Scolytidae) in response to chemical messengers. *Can. Entomologist* **101**, 143.

Pitman, G. B., Kliefoth, R. A., and Vite, J. P. (1965). Studies on the pheromone of *Ips confusus,* II. Further observations on the site of production. *Contrib. Boyce Thompson Inst.* **23**, 13.

Pitman, G. B., Renwick, J. A. A., and Vite, J. P. (1966). Studies on the pheromone of *Ips confusus* (Le Conte). IV. Isolation of the attractive substance by gas-liquid chromatography. *Contrib. Boyce Thompson Inst.* **25**, 243.

Pitman, G. B., Vite, J. P., Kinzer, G. W., and Fentiman, A. F., Jr. (1968). Bark beetle attractants: *trans*-verbenol isolated from *Dendroctonus. Nature* **218**, 168.

Pitman, G. B., Vite, J. P., Kinzer, G. W., and Fentiman, A. F., Jr. (1969). Specificity of population aggregating pheromones in *Dendroctonus. J. Insect Physiol.* **15**, 363.

Reece, C. A., Rodin, J. O., Brownlee, R. G., Duncan, W. G., and Silverstein, R. M. (1968). Syntheses of the principal components of the sex attractant from male *Ips confusus* Tetrahedron **24**, 4249.

Renwick, J. A. A. (1966). Chemical studies on the pheromone of *Ips confusus* (Coleoptera: Scolytidae). *Abstr. 16th Ann. Meeting Entomol. Soc. Canada, Banff, September, 1966.*

Renwick, J. A. A. (1967). Identification of two oxygenated terpenes from the bark beetles *Dendroctonus frontalis* and *Dendroctonus brevicomis. Contrib. Boyce Thompson Inst.* **23**, 355.

Renwick, J. A. A., and Vite, J. P. (1968). Isolation of the population aggregating pheromone of the southern pine beetle. *Contrib. Boyce Thompson Inst.* **24**, 65.

Rodin, J. O., Silverstein, R. M., Burkholder, W. E., and Gorman, J. E. (1969). Sex attractant of *Trogoderma inclusum. Science* **165**, 904.

Sci. Am. **215**, 64 (1966). The seduction of bark beetles.

Silverstein, R. M., and Bassler, G. C. (1967). "Spectrometric Identification of Organic Compounds," 2nd edition. Wiley, New York.

Silverstein, R. M., Rodin, J. O., and Wood, D. L. (1966a). Isolation and identification of the principal sex attractant components from the frass of *Ips confusus. Abstr. 16th Ann. Meeting Entomol. Soc. Canada, Banff, September, 1966.*

Silverstein, R. M., Rodin, J. O., and Wood, D. L. (1966b). Sex attractants in frass produced by male *Ips confusus* in ponderosa pine. *Science* **154**, 509.

Silverstein, R. M., Rodin, J. O., and Wood, D. L. (1967a). Methodology for isolation and identification of insect pheromones with reference to studies on the California five-spined *ips. J. Econ. Entomol.* **60**, 944.

Silverstein, R. M., Rodin, J. O., Burkholder, W. E., and Gorman, J. E. (1967b). Sex attractant of the black carpet beetle. *Science* **157**, 85.

Silverstein, R. M., Brownlee, R. G., Bellas, T. E., Wood, D. L., and Browne, L. E. (1968). Brevicomin: Principal sex attractant in the frass of the female western pine beetle. *Science* **159**, 889.

Vick, K. W., Burkholder, W. E., and Gorman, J. E. (1969). Interspecific response to sex pheromone of *Trogoderma species. Ann. Entomol. Soc. Am.* In press.

Vite, J. P. (1967). Sex attractants in frass from bark beetles. *Science* **155**, 105.

Vite, J. P. (1968). Timber industry attacks bark beetle problems through basic research. Texas Industry **36**, 14 (September 1968).

Vite, J. P., and Crozier, R. G. (1968). Studies on the attack behavior of the southern pine beetle. IV. Influence of host condition on aggregation pattern. *Contrib. Boyce Thompson Inst.* **24**, 87.

Vite, J. P., and Pitman, G. B. (1969). Insect and host odors in the aggregation of the western pine beetle. *Can. Entomologist* **101**, 113.

Wasserman, H. H., and Barber, E. H. (1969). Carbonyl-epoxide rearrangements. Synthesis of brevicomin and related [3.2.1]bicyclic systems. *J. Am. Chem. Soc.* **91**, 3674.

Wood, D. L., *et al.* In preparation.

Wood, D. L. (1962). The attraction created by males of a bark beetle, *Ips confusus* (LeC.) attacking ponderosa pine. *Pan-Pacific Entomologist* **38**, 141.

Wood, D. L., Browne, L. E., Silverstein, R. M., and Rodin, J. O. (1966). Sex pheromone of bark beetles. Mass production, bioassay, source, and isolation of the sex pheromone of *Ips confusus* (LeConte). *J. Insect Physiol.* **12**, 523.

Wood, D. L., Silverstein, R. M., and Rodin, J. O. (1967a). Sex attractants in frass from bark beetles. *Science* **155**, 105.

Wood, D. L., Stark, R. W., Silverstein, R. M., and Rodin, J. O. (1967b). Unique synergistic effects produced by the principal sex attractant compounds of *Ips confusus* (LeConte) (Coleoptera: Scolytidae). *Nature* **215,** 206.

Wood, D. L., Browne, L. E., Bedard, W. D., Tilden, P. E., Silverstein, R. M., and Rodin, J. O. (1968). Response of *Ips confusus* to synthetic sex pheromones in nature. *Science* **159,** 1373.

THE BOLL WEEVIL SEX ATTRACTANT*†‡

J. H. Tumlinson, R. C. Gueldner, D. D. Hardee,
A. C. Thompson, P. A. Hedin,
and J. P. Minyard

I. INTRODUCTION

The recent concern over the pollution of our environment and the disturbance of the ecological balance in nature owing to the use of insecticides has stimulated research in a variety of areas designed to control insects without using generally toxic materials. Interest in these investigations has also been enhanced by the rapid development of resistance to insecticides by many species.

Research at the Boll Weevil Research Laboratory has been designed to find noninsecticidal methods for the control or eradication of the boll weevil *Anthonomus grandis* Boheman. This insect is the major pest of cotton and one of the most important insects in terms of economic loss in the United States. Several chemical schemes intended

---

*Coleoptera: Curculionidae.

†In cooperation with Mississippi Agricultural Experiment Station. Part of a thesis submitted by J. H. Tumlinson in partial fulfillment of requirements for the Ph.D. degree from Mississippi State University, State College, Mississippi.

‡Mention of a proprietary product does not necessarily imply the endorsement of these products by the U.S. Department of Agriculture.

to affect the behavior and/or physiology of the weevil have been studied and are still being investigated.

That certain chemicals inhibit ovarian development and affect reproduction has been known for some time. The earlier literature on insect chemosterilants and on sterilization of the boll weevil was reviewed by Hedin *et al.* (1964, 1967). Studies at this laboratory (Hedin *et al.*, 1967) showed that the male boll weevil can be sterilized either by feeding or injecting both tepa [tris(1-aziridinyl)phosphine oxide] and apholate [2,2,4,4,6,6-hexakis(1-aziridinyl)-2,2,4,4,6,6-hexahydro-1,3,5,-2,4,6-triazatriphosphorine], but that mortality was high. At marginal levels of treatment about half the males eventually regained a degree of fertility, and this recovery was more rapid when insects were fed fresh cotton buds than when they were fed artificial diet. Investigations are continuing to find a material and method for effective large-scale sterilization of this insect.

The observation by Keller *et al.* (1962) that aqueous cotton bud extracts stimulated feeding by the boll weevil precipitated investigations into the feeding stimulant complex of the cotton bud. Pertinent literature on this subject was reviewed by Hedin *et al.* (1966). Isolational studies in the past 6 years (Hedin *et al.*, 1966; Struck *et al.*, 1968a, b; Temple *et al.*, 1968) indicated that no single compound in the plant evoked a full feeding stimulant response in the insect. Feeding activity is elicited by both polar and nonpolar solvent extracts. Subsequent fractionation implicated several classes of compounds, but activity decreased or even disappeared when the pure components were isolated. The most significant recent development was the formulation of a highly active feeding stimulant mixture for the boll weevil from known cotton constituents, common metabolites, and compounds inducing primary mammalian sensations of taste and odor (Hedin *et al.*, 1968).

Viehoever *et al.* (1918) first demonstrated that the volatile oil of Upland cotton was attractive to the boll weevil. More recently at our laboratory a search for the boll weevil plant attractant has been conducted in which the steam distillate of several tons of cotton buds and whole plants was fractionated and examined. The terpene and sesquiterpene hydrocarbons, the carbonyls, and the major volatile alcohol, β-bisabolol [1-(1,5-dimethyl-4-hexenyl)-4-methyl-3-cyclohexen-1-ol], were isolated, identified, and assayed for attractancy (Minyard *et al.*, 1965–1968). Bioassays were conducted as described by Hardee *et al.*, (1966). The terpene and sesquiterpene hydrocarbons are slightly repellent as a class, but α-pinene, limonene, and caryophyllene are attractive when assayed individually. β-Bisabolol is also attractive to weevils (60–80% of the total activity of the oil), as is caryophyllene oxide and two as yet unidentified sesquiterpenoid compounds present in very low concentrations (Minyard *et al.,* 1969). A number of compounds are thus involved and apparently they have some synergistic effect on each other when combined.

Insect hormones have recently been shown to have great potential for insect control. An investigation, underway at this laboratory, has shown juvenile hormone activity in boll weevil larval extracts (Gueldner, unpublished, 1970) and our study is progressing toward the isolation of the active components.

Investigations by workers at this laboratory (Keller *et al.,* 1964; Cross and Mitchell, 1966; Hardee *et al.,* 1967a,b) have shown that female boll weevils are responsive to a pheromone emitted by males. Hardee *et al.* (1967b) showed that 5-day-old males are the most attractive and 5-day-old females the most responsive. Additionally they showed that males fed cotton buds are more attractive to females than males fed artificial diet. Laboratory assays (Hardee *et al.,* 1967a) indicate that a true sex attractant is present since only females respond to males. Field tests, however, have shown that live caged males attract approximately equal numbers of males and females in the spring and fall, but mostly females in midseason (Cross and Hardee, 1968; Bradley *et al.,* 1968; Hardee *et al.,* 1969b). This may possibly be explained by a difference in the physiology of the overwintered and late season weevils, although confirming evidence is not yet available.

Comparison of the attractiveness of male weevils and cotton plants in the field indicates that males are considerably more attractive to males and females than is cotton (Hardee *et al.,* 1969b). Live males in field traps have been shown to be effective in controlling boll weevils only when populations of wild weevils are low (Hardee *et al.,* 1969a). Pheromone traps (using live males) in the Texas High Plains diapause control zone however, captured sufficient overwintered weevils to suppress the population until late summer migration overpowered the action of the traps (Hardee *et al.,* 1970).

Recent field investigations by Hardee suggest that control of a weevil population may be feasible using the sex pheromone in combination with other methods (chemosterilants, feeding stimulants, plant attractants, or combinations of all in insecticidal baits).

The promise of extraordinary practical value in using the boll weevil sex pheromone as a means of control, as compared to other methods, led us to initiate investigations in 1966 to determine the identity of the substances involved. The immediate goal of these studies was to isolate and determine the structures of the compounds responsible for the stimulus, and to confirm the postulated structures by synthesis. The ultimate aim, of course, is to produce, economically, synthetic compounds equal or superior to the natural attractants for use in eradication or control of the insect. We describe the achievement of this first goal in this paper.

II. APPARATUS AND MATERIALS

Carbowax 20M coated silica gel for use in liquid chromatography was prepared according to Kugler and Kovats (1963). Adsorbosil-CABN (Applied Science Labs) was washed with acetone–ether (1:1) and ether just prior to use in each liquid chromatographic column. A Barber-Colman Model 5000 unit and an Aerograph Model 600C chromatograph, both with flame ionization detectors, were used for the analytical gas chromatography. An Aerograph Autoprep Model A-700 with a thermal conductivity detector was used for all preparative work. The columns and conditions employed are shown in Table I. A hydrogenation apparatus similar to that described by Beroza and Sarmiento (1966) was used in conjunction with a Perkin-Elmer Model

TABLE I

Gas Chromatographic Operating Conditions

	Analytical Carbowax 4000	Analytical SE-30
Detector	H_2 flame	H_2 flame
Column length (ft)	6	20
Column diameter (in)	1/8	1/8
Percent stationary phase (w/w)	28.5	10
Solid support	60- to 80-mesh Gas-Chrom P	60- to 80-mesh Gas-Chrom Q
Carrier gas flow rate, (ml/min)	N_2:15	N_2:50
Column temperature (°C)	135	160

270 mass spectrometer which includes a gas chromatographic inlet system. Nuclear magnetic resonance (NMR) spectra were obtained with a Varian A-60 analytical NMR spectrometer and infrared (IR) spectra with a Perkin-Elmer Model 521 infrared spectrophotometer. Olefins were ozonized with a Supelco Micro-Ozonizer (Beroza and Bierl, 1966, 1967).

III. PROCEDURE

Extractions and further purifications from fecal material and insects were carried out in identical fashion throughout this investigation. Each step was monitored by laboratory bioassay (Hardee *et al.,* 1967a) of individual fractions and combinations of various fractions. Insects (67,000 males and 4,500,000 weevils of mixed sexes) or fecal material (54.7 kg) were extracted with dichloromethane; the extract was concentrated under vacuum and steam-distilled. The distillate was extracted with dichloromethane and the solvent again removed under vacuum. The concentrated distillate extract was then chromatographed on a Carbowax 20M coated silica gel column which was eluted successively with pentane, pentane–ether (90:10), pentane–ether (50:50), ether, and methanol. The (90:10) and (50:50) fractions, which

(continued)

Analytical Carbowax 20 M	Preparative SE-30	Preparative SE-30
Mass spectrometer	Thermal conductor	Thermal conductor
50	20	20
0.02 (i.d.)	1/8	3/8
SCOT[a]	10	10
–	60- to 80-mesh Gas-Chrom Q	60- to 80-mesh Gas-Chrom P
He:2-3	He:20	He:200
110	160	160

[a] SCOT = Support Coated Open Tubular.

were active when combined, were further chromatographed individually on Adsorbosil-CABN (25% AgNO$_3$ on silica gel), and eluted with the same series of solvents except for methanol. For both the (90:10) and (50:50) silica gel fractions, the subfraction eluted from Adsorbosil with pentane–ether (50:50) was active when combined with its counterpart from the other fraction.

The material eluted with (90:10) pentane–ether from Carbowax 20M/silica gel and (50:50) pentane–ether from AgNO$_3$/silica gel gave six peaks on gas-liquid chromatography (GLC) on Carbowax 4000. Peak 6 gave two peaks, 6A and 6B, on Carbowax 20M.

The material eluted with (50:50) pentane–ether from both liquid columns gave eight peaks on GLC on Carbowax 4000. Peak 8 gave two peaks, 8A and 8B, on SE-30.

The structures of compounds 8A and 8B were deduced from their NMR and IR spectra, in addition to the mass spectra of the original compounds and products from catalytic reduction in the gas chromatograph-mass spectrometer. Compound 8B was ozonized (Beroza and Bierl, 1966, 1967) to confirm its hypothesized structure. Structures of 6A and 6B were deduced from their mass spectra and the mass spectrum of the single catalytic reduction product. Structures of these four compounds have been confirmed by unequivocal synthesis.

IV. RESULTS AND DISCUSSION

A. ISOLATION

Extracts and steam distillate extracts of male boll weevils were attractive to females in bioassays. Extracts of weevils of mixed sexes were not attractive to females, but steam distillation of this extract produced a material very attractive to females. Apparently steam distillation removed some masking agent. In no case were any of the extracts ever attractive to males in laboratory assays, nor were materials derived from females attractive to either sex.

The occurrence of sex pheromones in the frass or fecal material of male California five-spined ips, *Ips confusus* (LeConte), was reported by Wood *et al.* (1966). Steam distillation of the extract of fecal material of boll weevils (both male and mixed sexes) produced a material highly attractive to females but not to males (Tumlinson *et al.,* 1968). The ease of obtaining fairly large quantities of fecal material from the laboratory colony (about 200 gm per day from 200,000 insects) as a byproduct of rearing weevils made it a valuable source of the sex pheromone.

Chromatography of the active steam distillates on Carbowax 20M coated silica gel produced five fractions, none of which were attractive to females when assayed individually. Recombination of the various fractions showed that all the activity of the steam distillate was recovered when fractions 2 and 3 [eluted with (90:10) and (50:50) pentane–ether, respectively] were combined.

When these two active fractions were chromatographed individually on Adsorbosil-CABN they each gave four inactive fractions. Again various combinations showed that when the pentane–ether (50:50) eluate (fraction 23) of fraction 2 was combined with the pentane–ether (50:50) eluate (fraction 33) of fraction 3, the original activity was recovered.

When chromatographed on the analytical Carbowax 4000 column, fraction 33 gave eight peaks, the eighth peak constituting about 90% of the entire fraction. When these compounds were collected from the analytical column by means of a stream splitter, component 8 proved to be attractive to female weevils when it was combined with fraction 23. Gas-liquid chromatography of peak 8 on SE-30 gave two peaks, 8A and 8B. Neither of these components was active when individually combined with fraction 23 and assayed. Combination of both components with fraction 23 produced a mixture as highly attractive to females as the original distillate. Component 8 had a retention index of 1820 on Carbowax 4000, and components 8A and 8B had retention indices of 1205 and 1228, respectively, on SE-30 (Kovats, 1961). Both components were collected in carbon tetrachloride for subsequent spectral studies.

B. IDENTIFICATION

The significant peaks in the mass spectrum of component 8A were a parent peak at m/e 154; 139, loss of $-CH_3$; 136, loss of H_2O; 109, loss of $-CH_2-CH_2OH$; and 68, base peak, resulting from ring cleavage of the dehydrated ion from compound (I) to produce a charged isoprene fragment. This evidence was supported by degradation of 8A on attempted catalytic reduction (140°C) in the gas chromatograph, as would be expected of a cyclobutane ring. The infrared spectrum in CCl_4 showed a sharp signal at 3630 cm^{-1} (free OH stretch) and a broad hydrogen-bonded hydroxyl signal from about 3250 to 3550 cm^{-1}; strong signals at 885 and 1642 cm^{-1} suggested a terminal methylene. The structure of 2-isopropenyl-1-methylcyclobutaneethanol (I) for compound 8A was demanded by the NMR spectrum. The upfield shift of a one proton signal (2.59 ppm, δ) on dilution and the disappearance of this signal on the addition of trichloroacetyl isocyanate (TCAIC) confirm the -OH group. Two broadened signals at 4.71 ppm (one proton) and 4.88 ppm (one proton) are evidence of a terminal methylene with no protons on the adjacent carbon of the double bond. Together with the vinyl methyl singlet at 1.72 ppm, this strongly supports the existence of an isopropenyl group. This single unsaturation also requires a monocyclic structure. A two-proton triplet at 3.63 ppm ($-CH_2-CH_2-OH$) suggests a primary alcohol on a chain consisting of at least two methylenes. A broadened triplet at 2.60 ppm (one methinyl proton adjacent to a methylene) and the very sharp singlet at 1.22 ppm (three protons) suggesting a methyl attached to a tetrasubstituted carbon, together with the broad multiplet from about 1.3 to 2.2 ppm (six protons), support the structure assigned to compound (I) (8A). Two of the latter six protons have been accounted for ($-CH_2-CH_2-OH$) already. The possibility of 1,3-substitution on the cyclobutane ring rather than 1,2-substitution cannot be ruled out completely; however, this would violate the isoprene rule.

(I)   (II)   (III)   (IV)

Compound 8B was identified as *cis*-3,3-dimethyl-Δ$^{1,\beta}$-cyclohexaneethanol (II) on the basis of its mass, NMR, and IR spectra. The mass spectrum showed a parent peak at *m/e* 154, and a peak at *m/e* 136, attributed to the loss of H_2O by the parent as would be expected of an alcohol. Other significant peaks were: *m/e* 121 (P-33), loss of -CH$_3$ and

H_2O; 107, loss of CH_3OH and $-CH_3$; and the base peak at m/e 69 attributable to cleavage of the six-membered ring allylic to the double bond in structure (II) to yield $CH_3C(CH_3)=CHCH_2^+$.

Production of the m/e 107 fragment would be particularly favored in the cis structure shown for 8B via a six-membered cyclic rearrangement involving a proton transfer, loss of the elements of methanol, and methyl radical expulsion as shown below.

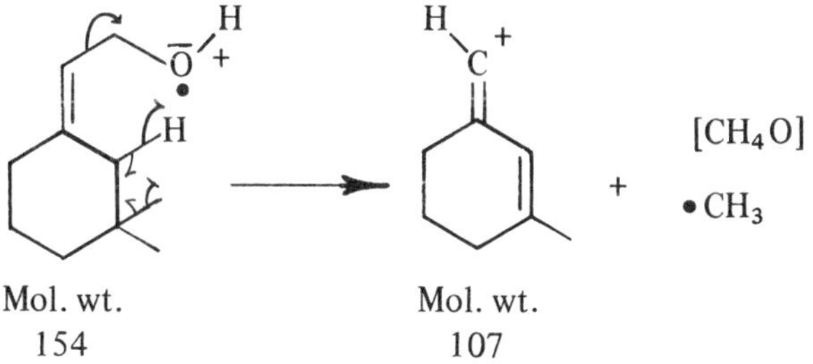

Mol. wt.	Mol. wt.
154	107

Reduction of compound 8B on a palladium catalyst immediately ahead of a Carbowax 20M SCOT column gave a compound with a parent mass of 156 and other major fragments in the spectrum two mass units higher than the corresponding peaks in 8B.

The OH stretch at 3610 cm⁻¹, less hydrogen-bonded than in compound (I), the upfield shift of one proton signal (1.90 ppm, δ) in the NMR spectrum on dilution, and removal of this signal on addition of TCAIC confirm that the compound is an alcohol. The following NMR signals support the assigned structure, (II): triplet, 5.53 ppm (one proton, J = 7.0 cps, RR′C = CH–CH_2OH), and doublet, 4.05 ppm (two protons, J = 7.0 cps, RR′C = CH–CH_2-OH) shifted downfield by addition of TCAIC; a prominent singlet, 2.00 ppm (two protons) overlapping a multiplet at about 2.09 ppm (two protons) suggesting two methylenes adjacent to a double bond, one split and the other not; broad multiplet, 1.13 to 1.83 ppm (four protons R–CH_2CH_2-R′); and a sharp singlet, 0.95 ppm (six protons, geminal dimethyls). The cis configuration about the double bond was assigned by comparison of the NMR spectra of the cis and trans synthetic ester precursors (see below).

Further proof of structure for compound (II) was obtained by microozonolysis. One major component was obtained which had NMR spectra, mass spectra, and GLC behavior identical to 3,3-dimethylcyclohexanone.

Chromatography of fraction 23 on the analytical Carbowax 4000 column produced six peaks. When they were assayed in combination with fraction 33, component 6 proved to be the only active one. Chromatography of component 6 on the analytical SE-30 column gave only one peak which retained its activity in the bioassays. Component 6 had a retention index of 1782 on Carbowax 4000 and 1248 on SE-30. When eluted from the gas chromatograph into (2,4-dinitrophenyl)hydrazine reagent on a thin-layer plate (Tumlinson *et al.,* 1967), it produced a derivative which had an R_f similar to that of standard terpene carbonyls. GLC of component 6 on the Carbowax 20M SCOT column gave two components, 6A and 6B. The mass spectra of these two compounds were nearly identical to each other and similar to compound 8B. The parent peak had a *m/e* of 152 in both cases, appropriate for a monocyclic terpene aldehyde or ketone with one unsatured bond. Reduction of 6A and 6B at the inlet of the gas chromatograph equipped with the SCOT column in the manner of Beroza and Sarmiento (1966) produced only one peak with a mass of 154, confirming the single unsaturated bonds. The base peak in the saturated 6A and 6B spectrum, *m/e* 110, suggests a facile loss of the elements of acetaldehyde. Such a rearrangement peak suggested a $-CH_2-CH=O$ side chain that might easily cleave by a cyclic rearrangement process analogous to the following.

Mol. wt.	Mol. wt.	Mol. wt.
154	110	44

On the basis of these data, structures (III) (*cis*-3,3-dimethyl-$\Delta^{1,\alpha}$-cyclohexaneacetaldehyde) and (IV) (*trans*-3,3-dimethyl-$\Delta^{1,\alpha}$-cyclohexaneacetaldehyde), the aldehydic analogs of compound (II) (8B), were postulated for 6A and 6B respectively. These structures were quickly confirmed by synthesis (see below).

Compounds 6A and 6B were not assayed separately for attractiveness, but always together as component 6 collected from an SE-30 GLC column. Assays showed that compounds 8A, 8B, and 6 must all be present to attract female weevils. Absence of any one renders the mixture almost completely unattractive.

C. QUANTITATIVE ESTIMATES

The amount of each compound present was estimated by addition of a known amount of a standard compound to fractions 23 and 33, gas chromatography of these fractions, and measurement of the peak areas of the standards and natural components. The quantity of fecal material or insects from which each fraction was derived was used to estimate the concentration of each component in the starting material. Compounds, 8A, 8B, 6A, and 6B are present in concentrations of 0.76, 0.57, 0.06, and 0.06 ppm, respectively, in fecal material. Concentrations in mixed weevils are about tenfold less. One insect produces about 1 mg of fecal material per day. Estimating that 50% of the weevils were males, one male equivalent of material was calculated to be that amount derived from 2 mg of fecal material from weevils of mixed sexes. Bioassays over a range of concentrations of compounds 8A, 8B, and 6 showed that with materials derived from fecal material, 50 male equivalents each of 8A and 8B combined with about 2000 male equivalents of 6 gave optimum activity. Concentrations from about 5 to 1000 male equivalents of 8A or 8B and from 200 to 10,000 male equivalents of 6 retained above 50% of optimum activity.

D. SYNTHESIS

Compound (I), its trans isomer, and the 1,3-substituted cyclobutane isomers were synthesized by the sequence shown in Fig. 1. The photocycloaddition produced several other products besides those shown in Fig. 1. The GLC behavior and spectral data for synthetic *cis-*2-isopropenyl-1-methylcyclobutaneethanol were identical to those of natural compound (I) (8A). Synthetic *trans*-(I) had almost identical GLC behavior to the cis isomer. Its NMR spectrum was also very similar except for the upfield shift of the methyl singlet from 1.22 ppm in the cis isomer to about 0.95 ppm in the trans isomer. The trans isomer was active in laboratory assays at concentrations 100- to 200-fold greater than the cis.

The scheme in Fig. 2 shows the synthesis of compound (II) (8B). 3,3-Dimethylcyclohexanone was produced from 3-methyl-2-cyclohexen-

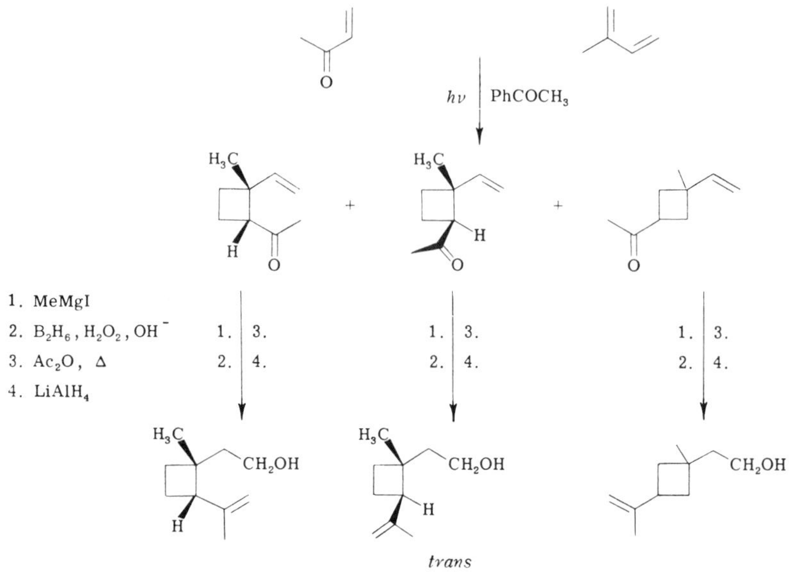

FIG. 1.   Synthesis scheme for compound (I).

l-one by the method of Buchi *et al.* (1948). A Reformatsky reaction of ethyl bromoacetate with the dimethylcyclohexanone followed by saponification then yielded l-hydroxy-3,3-dimethylcyclohexaneacetic acid (75%). The hydroxy acid was dehydrated with acetic anhydride (Elliott and Linstead, 1938) and the unsaturated acids esterified with ethanol. The cis and trans unsaturated esters were separated by GLC on the preparative SE-30 column and reduced to the respective cis and trans alcohols with LiAlH$_4$.

Assignment of the cis and trans configurations was made by comparing the NMR spectra of the cis and trans esters. The spectrum of the cis ester showed a singlet at 5.72 ppm (one proton, olefinic) and a singlet at 2.72 ppm (two protons, EtOOC–CH = CR–CH_2–CR′,R″, R‴). This latter signal indicated a considerable paramagnetic deshielding of the methylene at the 2-position on the ring, by the spatially adjacent carbonyl group in the cis ester. The trans ester, on the other hand, had a singlet at 5.58 ppm (one proton, olefinic) produced by the more shielded proton now cis to the geminal methyls. A triplet at 2.85 ppm (two protons) is due to the ring methylene at the 6-position which

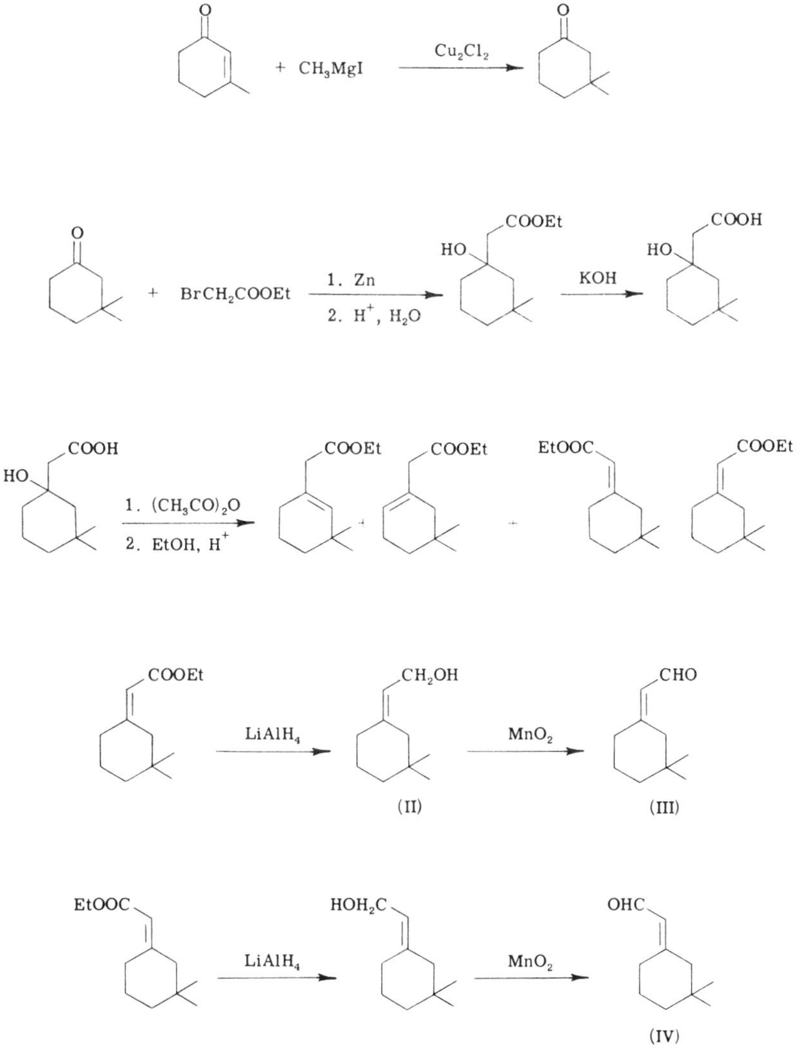

FIG. 2. Synthesis scheme for compounds (II), (III), and (IV).

is now deshielded inductively by the adjacent cis carbethoxyl group and split by the adjacent ring methylene. The ring methylene at the 2-position gives rise to a singlet at 2.05 ppm, well upfield of the analogous methylene in the cis ester. These assignments are consistent with Bible (1968). The cis alcohol was identical in all respects, including insect attractancy, to peak 8B, substantiating structure (II).

The cis and trans alcohols were readily converted to the cis and trans aldehydes, respectively, by selective oxidation with active MnO_2 (Corey *et al.,* 1968). The mass spectrum and GLC retention behavior of the cis aldehyde were identical to that of 6A, whereas those of the trans aldehyde were identical to 6B. Thus, structures (III) and (IV) are correct for 6A and 6B, respectively.

The synthetic compounds (I–IV), are as attractive as their natural analogs in the bioassays when combined in the proper ratio with the other components, either synthetic or natural. To our knowledge none of these four compounds have been found previously in natural products nor have they been synthesized and characterized. It seems likely that compound (II) was synthesized by Daessle and Schinz (1956), enroute to the synthesis of 3-methyl-4-allocyclogeranyl-2-butenoic acid, but it was not isolated and characterized.

As previously mentioned, production of the sex pheromone appears to be diet-related, since fresh cotton squares produce more attractive males than artificial diet. This leads to the postulation that the weevil may be converting a plant constituent into the sexually attractive compounds. This is even more appealing when it is considered that monoterpene alcohols or pyrophosphate ester precursors of both *trans-β*-ocimene and myrcene are expected in the plant since these two hydrocarbons are principal terpenes in the essential oil of the cotton bud (Minyard *et al.,* 1965). Figure 3 shows a postulated pathway by which a terpene alcohol could be converted to all four active compounds. We hasten to add that this is purely speculative and that we have no supporting evidence at this time. Considering the rather unusual structures of the active compounds and the related circumstances, however, we think this proposal very reasonable.

The attractiveness of male weevils to both sexes in the field in early spring and fall, and to females only in the field in midsummer and in the laboratory at all times is subject to at least two interpretations. A physiological or dietary change in either or both sexes may be involved, with the four components adequate for evoking field aggregation and/ or sex attractancy responses. Alternatively, additional compounds may be required for *(a)* acting synergistically with the four compounds to produce either or both responses in the field or *(b)* acting separately from the four for sex attractancy or evoking the aggregating response without being involved in laboratory sex attractancy.

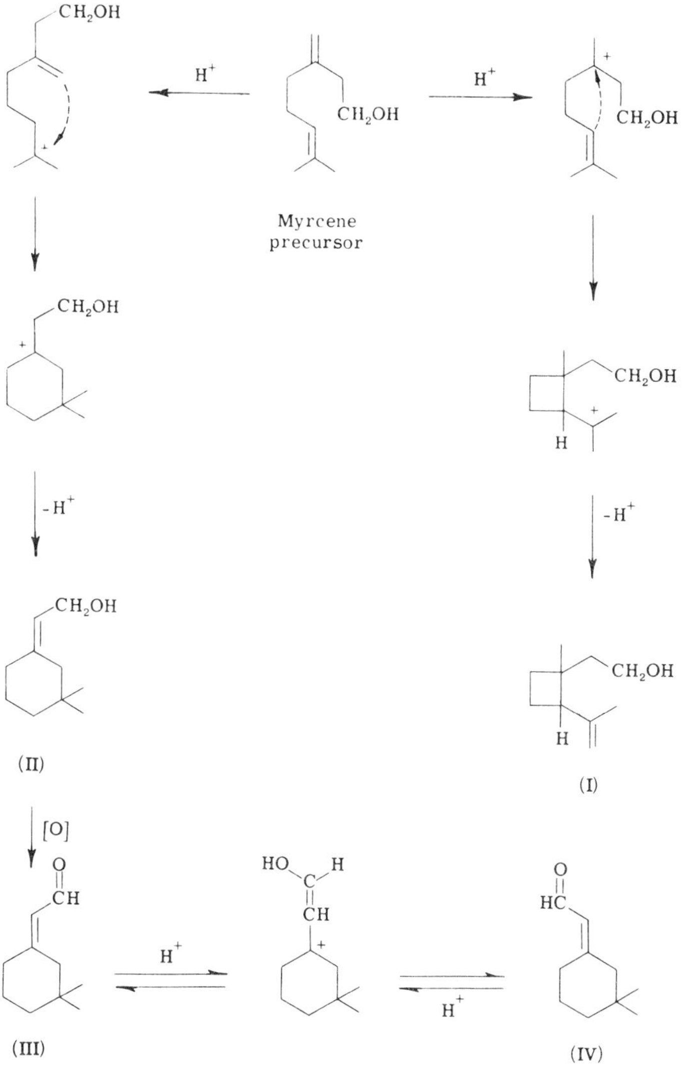

FIG. 3. Hypothetical biosynthetic scheme.

An example of the requirement for an additional compound to attract one sex only is the pink bollworm moth *Pectinophora gossypiella* (Saunders), in which the laboratory sex attractant, 10-propyl-*trans*-5,9-tridecadienyl acetate, emitted by females to attract males, required a synergist, *N,N*-diethyl-*m*-toluamide to attract males in the field (Jones *et al.*, 1966; Jones and Jacobson, 1968). Indeed this may be the case for several insects for which one compound has been identified as the sex attractant on the basis of laboratory bioassays, but which has met with no great success in field tests. In another instance, with *Ips confusus* (LeConte), more than one compound was required for full response in the laboratory and the synthetic pheromone mixture produced full response in field tests (Silverstein *et al.*, 1966a,b; Wood *et al.*, 1968).

Since *Anthonomus grandis* Boheman and *Ips confusus* (LeConte) stem from closely related families, and since both require several components of similar structure, there is precedent to expect that the three or four compounds sufficient for attraction of females under laboratory conditions will also attract females and possibly both sexes in the field. Laboratory males fed fresh cotton evoked a pure sex attractant response in the laboratory bioassay and an aggregating response in the field. Therefore we might hope that the four compounds will produce both sex attractant and aggregant responses at appropriate seasons in the field. The ability of these compounds to elicit an oriented sex-specific insect movement over an appreciable distance in the laboratory bioassay would appear to support at least the expectation of field sex attractant activity for this mixture.

In all laboratory bioassays conducted so far, only laboratory-reared weevils or wild midsummer males have been used. Only males have been attractive, and these to females. Entomologists at our laboratory plan to determine whether an aggregating response can be obtained with field-overwintered insects in the laboratory. They also plan field tests with the natural and synthetic pheromones in the spring and summer of 1969.

SUMMARY

The isolation and identification of four terpenoid compounds from the male boll weevil, which, when properly combined, attract the female, are reported. The structures of all four have been confirmed by synthesis and each is biologically active.

The active compounds were isolated from both fecal material and insects by steam distillation and a series of chromatographic separations. They were identified, on the basis of chemical and spectral evidence, as: (I), *cis*-2-isopropenyl-l-methylcyclobutaneethanol; (II), *cis*-

3,3-dimethyl-Δ $^{1,\beta}$-cyclohexaneethanol; (III), *cis*-3,3-dimethyl-Δ $^{1,\alpha}$-cyclohexaneacetaldehyde; (IV), *trans*-3,3-dimethyl-Δ $^{1,\alpha}$-cyclohexaneacetaldehyde. These compounds were present in fecal material in concentrations of 0.76, 0.57, 0.06, and 0.06 ppm, respectively. Compounds (I)–(IV) have been synthesized and the synthetic compounds are equally as attractive as the natural substances to females.

To produce a substance attractive to females in laboratory bioassays it was necessary to combine compounds (I), (II), and either (III) or (IV). Absence of either of the alcohols or both of the aldehydes resulted in an inactive mixture. These compounds have been assayed only under laboratory conditions, but extracts from male weevils in earlier tests showed activity in field studies. Live male boll weevils attract approximately equal numbers of males and females in the field in spring and fall, indicating that these compounds may act as an aggregating pheromone as well as a sex pheromone.

A number of noninsecticidal chemical methods for controlling the boll weevil (chemosterilants, plant attractants, feeding stimulants, and hormones), which have or are being investigated have been reviewed.

Note added in proof:
The results of this investigation were reported by Tumlinson *et al.* (1969). In field tests (late 1969 and early 1970) the synthetic mixture (I–IV) was found to be highly competitive with live male weevils. Two new syntheses of compound (I) were announced in *Chemical and Engineering News,* January 26, 1970, pages 40 and 43.

REFERENCES

Beroza, M., and Bierl, B. A. (1966). Apparatus for ozonolysis of microgram to milligram amounts of compound. *Anal. Chem.* **38**, 1976.

Beroza, M., and Bierl, B. A. (1967). Rapid determination of olefin position in organic compounds in microgram range by ozonolysis and gas chromatography. *Anal. Chem.* **39**, 1131.

Beroza, M., and Sarmiento, R. (1966). Apparatus for reaction chromatography. Instantaneous hydrogenation of unsaturated esters, alcohols, ethers, ketones, and other compound types and determination of their separation factors. *Anal. Chem.* **38**, 1042.

Bible, R. H., Jr. (1968). "Interpretation of NMR Spectra, an Empirical Approach," pp. 17-18. Plenum Press, New York.

Bradley, J. R., Clower, D. F., and Graves, J. B. (1968). Field studies of sex attraction in the boll weevil. *J. Econ. Entomol.* **61**, 1457.

Buchi, G., Jeger, O. and Ruzicka, L. (1948). Zur Kenntnis der Triterpene. Synthese des $\Delta^{5,10}$-1, 1-Dimethyl-octalons-(6), eines Abbauproduktes des Ambreins. *Helv. Chim. Acta* **31**, 241.

Corey, E. J., Gilman, N. W. and Ganem, B. E. (1968). New methods for the oxidation of aldehydes to carboxylic acids and esters. *J. Am. Chem. Soc.* **90**, 5616.

Cross, W. H., and Hardee, D. D. (1968). Traps for survey of overwintered boll weevil populations. *Coop. Econ. Inst. Rept.* **18**, 430.

Cross, W. H., and Mitchell, H. C. (1966). Mating behavior of the female boll weevil. *J. Econ. Entomol.* **59**, 1503.

Cross, W. H., Hardee, D. D., Nichols, F., Mitchell, H. C., Mitchell, E. B., Huddleston, P. M., and Tumlinson, J. H. (1969). Capture of female boll weevils in sex attractant traps containing males or male extract. *J. Econ. Entomol.* **62**, 154.

Daessle, C. L., and Schinz, H. (1956). Preparation et cyclisation du methyl-3-allo cyclogeranyl-4-butene-2-oique. *Helv. Chim. Acta* **39**, 2118.

Elliott, G. H., and Linstead, R. P. (1938). 2,2-Dimethylcyclohexylacetic acid. *J. Chem. Soc.* p.776.

Hardee, D. D., Mitchell, E. B., Huddleston, P. M., and Davich, T. B. (1966. A laboratory technique for bioassay of plant attractants for the boll weevil. *J. Econ. Entomol.* **59**, 240.

Hardee, D. D., Mitchell, E. B., and Huddleston, P. M. (1967). Procedure for bioassaying the sex attractant of the boll weevil. *J. Econ. Entomol.* **60**, 169. Hardee, D. D., Mitchell, E. B., and Huddleston, P. M. (1967b). Laboratory studies of sex attraction in the boll weevil. *J. Econ. Entomol.* **60**, 1221.

Hardee, D. D., Cross, W. H., Mitchell, E. B., Huddleston, P. M., Mitchell, H. C., Merkl, M. E., and Davich, T. B. (1969a). Biological factors influencing responses of the female boll weevil to the male sex pheromone in field and largecage tests. *J. Econ. Entomol.* **62**, 161.

Hardee, D. D., Cross, W. H., and Mitchell, E. B. (1969b). Male boll weevils are more attractive than cotton plants to boll weevils. *J. Econ. Entomol.* **62**, 165.

Hardee, D. D., Cross, W. H., Huddleston, P. M., and Davich, T. B. (1970). Survey and control of the boll weevil in west Texas with traps baited with males. *J. Econ. Entomol.*, in press.

Hedin, P. A., Cody, C. P., and Thompson, A. C. (1964). Antifertility effect of the chemosterilant apholate on the male boll weevil. *J. Econ. Entomol.* **57**, 270.

Hedin, P. A., Thompson, A. C., and Minyard, J. P. (1966). Constituents of the cotton bud. III. Factors that stimulate feeding by the boll weevil. *J. Econ. Entomol.* **59**, 181.

Hedin, P. A., Wiygul, G., Vickers, D., Bartlett, A., and Mitlin, N. (1967). Sterility induced by tepa in the boll weevil: effective dose and permanency, gonadal changes, and biological turnover of labeled compound. *J. Econ. Entomol.* **60**, 209.

Hedin, P. A., Miles, L. R., Thompson, A. C., and Minyard, J. P. (1968). Constituents of a cotton bud. Formulation of a boll weevil feeding stimulant mixture. *J. Agr. Food Chem.* **16**, 505.

Jacobson, M. (1965). "Insect Sex Attractants." Wiley (Interscience), New York.

Jones, W. A., and Jacobson, M. (1968). Isolation of *N,N*-Diethyl-*m*-toluamide (Deet) from female pink bollworm moths. *Science* **159**, 99.

Jones, W. A., Jacobson, M. and Martin, D. F. (1966). Sex attractant of the pink bollworm moth: Isolation, identification, and synthesis. *Sciencee* **152**, 1516.

Keller, J. C., Maxwell, F. G., and Jenkins, J. N. (1962). Cotton extracts as arrestants and feeding stimulants for the boll weevil. *J. Econ. Entomol.* **55**, 800.

Keller, J. C., Mitchell, E. B., McKibben, G., and Davich, T. B. (1964). A sex attractant for female boll weevils from males. *J. Econ. Entomol.* **57**, 609.

Kovats, E. (1961). Zusammenhange zwischen struktur and gaschromatographischen Daten organishcher verbindungen. *Z. Anal. Chem.* **181**, 351.

Kuglar, E., and Kovats, E. (1963). Zur Kenntnis des mandarinen-schalen Ols (*Citrus reticulata* Blanco, bzw. *Citrus nobilis* var. *deliciosa* Swingle; "Mandarin"). *Helv. Chim. Acta* **45**, 1480.

Minyard, J. P., Tumlinson, J. H., Hedin, P. A. and Thompson, A. C. (1965). Constituents of the cotton bud. Terpene hydrocarbons. *J. Agr. Food Chem.* **13**, 599.

Minyard, J. P., Tumlinson, J. H., Hedin, P. A., and Thompson, A. C. (1966). Constituents of the cotton bud. Sesquiterpene hydrocarbons. *J. Agr. Food Chem.* **14**, 332.

Minyard, J. P., Tumlinson, J. H., Hedin, P. A., and Thompson, A. C. (1967). Constituents of the cotton bud. The carbonyl compounds. *J. Agr. Food Chem.* **15**, 517.

Minyard, J. P., Thompson, A. C., and Hedin, P. A. (1968). Constituents of the cotton bud. VIII. β-Bisabolol, a new sesquiterpene alcohol. *J. Org. Chem.* **33**, 909.

Minyard, J. P., Hardee, D. D., Gueldner, R. C., Thompson, A. C., Wiygul, G., and Hedin, P. A. (1969). Constituents of the cotton bud. Compounds attractive to the boll weevil. *J. Agr. Food Chem.* **17**, 1093.

Silverstein, R. M., Rodin, J. O., and Wood, D. L. (1966a). Sex attractants in frass produced by male *Ips confusus* in ponderosa pine. *Science* **154**, 153.

Silverstein, R. M., Rodin, J. O., Wood, D. L., and Browne, L. E. (1966b). Identification of two new terpene alcohols from frass produced by *Ips confusus* in ponderosa pine. *Tetrahedronn* **22**, 1929.

Struck, R. F., Frye, J., Shealy, Y. F., Hedin, P. A., Thompson, A. C., and Minyard, J. P. (1968a). Constituents of the cotton bud. IX. Further studies on a polar boll weevil feeding stimulant complex. *J. Econ. Entomol.* **61**, 270.

Struck, R. F., Frye, J., Shealy, Y. F., Hedin, P. A., Thompson, A. C., and Minyard, J. P. (1968b). Constituents of the cotton bud. XI. Studies of a feeding stimulant complex from flower petals for the boll weevil. *J. Econ. Entomol.* **61**, 664.

Temple, C., Roberts, E. C., Frye, J., Struck, R. F., Shealy, Y. F., Thompson, A. C., Minyard, J. P., and Hedin, P. A. (1968). Constituents of the cotton bud. XIII. Further studies on a nonpolar feeding stimulant for the boll weevil. *J. Econ. Entomol.* **61**, 1388.

Tumlinson, J. H., Minyard, J. P., Hedin, P. A., and Thompson, A. C. (1967). Reaction chromatography I. Gas-liquid/thin-layer chromatographic derivatization technique for the identification of carbonyl compounds. *J. Chromatog.* **29**, 80.

Tumlinson, J. H., Hardee, D. D., Minyard, J. P., Thompson, A. C., Gast, R. T., and Hedin, P. A. (1968). Boll weevil sex attractant: Isolation studies. *J. Econ. Entomol.* **61**, 470.

Tumlinson, J. H., Hardee, D. D., Gueldner, R. C., Thompson, A. C., Hedin, P. A., and Minyard, J. P. (1969). Sex pheromones produced by the male boll weevil: isolation, identification, and synthesis. *Science,* **166**, 1010.

Wood, D. L., Browne, L. E., Silverstein, R. M., and Rodin, R. O. (1966). Sex pheromones of bark beetles - I. Mass production, bio-assay, source, and isolation of the sex pheromone of *Ips confusus* (LeConte). *J. Insect Physiol.* **12**, 523.

Wood, D. L., Browne, L. E., Bedard, W. D., Tilden, P. E., Silverstein, R. M., and Rodin, J. O. (1968). Response of *Ips confusus* to synthetic sex pheromones in nature. *Science* **159**, 1373.

Viehoever, A., Chernoff, L. H., and Johns, C. O. (1918). Chemistry of the cotton plant with special reference to upland cotton. *J. Agr. Res.* **13**, 345.

THE CHEMICAL BASIS OF
INSECT SOCIALITY

Murray S. Blum

I. INTRODUCTION

Among invertebrate animals, sociality reaches its highest degree of development in the Hymenoptera and Isoptera. Thus, among species of bees, wasps, ants, and termites, the large populations often present in a single colony appear to be characterized by a high degree of social coordination and interaction. It is obvious that, as in higher animals, these insects have developed effective means to integrate the functions of large populations in order to permit sociality to become a *fait accompli*. It has only been in the last two decades, however, that the nature of these integrative mechanisms has been elucidated as a primary requisite to understanding how sociality developed in the invertebrates. Although it would be premature to conclude at this juncture that all elements which regulate the interplay of individuals in

61

the social milieu have been illuminated, it is nevertheless possible now to examine critically the behavior of social insects in terms of some concrete control mechanisms.

In a sense, our comprehension of social regulators has paralleled the decipherment of the communication codes which are employed by hymenopterous and isopterous insects. The social cohesiveness of an insect colony appears to be maintained primarily through the utilization of a series of communication chemicals which function to transmit specific bits of information to members of a single population. These chemicals, the pheromones, are produced in specialized tissues called exocrine glands which evacuate their contents into the environment in response to specific stimuli. Social insects possess a multitude of exocrine glands (Pavan and Ronchetti, 1955), whose functions have yet to be elucidated, but it seems likely that the subtleties of insect sociality will become more comprehensible in proportion to our degree of understanding of the exact roles of the exocrine secretions. At the present time, it is convenient to examine chemical sociality in terms of single pheromones whose information content is rapidly decoded by the target insects in a population. It is not unlikely, however, that the simultaneous secretion from two or more exocrine organs by the transmitting insect may enable the receiver individuals to gain increased communication efficacy by modifying the informational content which is carried by a single pheromone.

Comparative exocrinology is a new science, a development which underscores the fact that biologists have only recently become aware of the importance of chemical information transfer as a reliable means of communicating. When viewed in this light, it is not particularly surprising that only the most obvious classes of pheromones have been recognized and in some cases experimentally verified. The rapid increase in the number of exocrinological investigations which have characterized this past decade almost guarantees that pheromones with cryptic roles will be illuminated in the near future.

II. TRAIL PHEROMONES

Although the existence of odor trails in social insects has been recognized for years, only recently has it been possible to determine that odor trail pheromones originate in specific glandular tissues in ants

(Wilson, 1959), bees (Lindauer and Kerr, 1958), and termites (Stuart, 1961). Formicid trail pheromones have been examined in some detail in the last decade and some aspects of trail laying in ants have recently been reviewed (Gabba, 1967).

A. ANT TRAIL PHEROMONES

Ant trail pheromones originate in several different glandular tissues (Table I)—a fact which serves to emphasize the polyphyletic origin of trail laying in the family Formicidae. Odor trail pheromones apparently originate in the hind gut in three subfamilies of ants, but it has not been established unequivocally whether the odor trail pheromones represent digestive constituents rather than a glandular product elaborated from hind gut cells (Blum and Portocarrero, 1964). It is also possible that the trail pheromones present in the hind gut have their origins more anteriorly in the digestive tract.

Dolichoderine ants have developed a novel social organ, Pavan's gland, in which the trail pheromones probably are biosynthesized (Wilson and Pavan, 1959). The widespread occurrence of this organ among dolichoderine genera (Pavan, 1955) indicates that this exocrine structure is of major significance in the social biology of these ants.

It is only in the ant subfamily Myrmicinae that the glands associated with the sting have been utilized as the source of the trail pheromone. Wilson (1959) demonstrated that the contents of Dufour's gland of the fire ant *Solenopsis saevissima* are evacuated through the sting in order to generate odor trails. On the other hand, several myrmicine genera lay odor trails with the poison gland secretion itself, a novel function for the venomous natural products which heretofore had been considered solely as defensive products (Blum *et al.,* 1964). Ants in the myrmicine genus *Crematogaster,* however, pay out their trail pheromone through the tarsi of their metathoracic legs (Fletcher and Brand, 1968), an extraordinary deviation from the trail-laying gastric themes which are practiced by ants in all other genera for which the pheromone-producing organs have been identified. The pheromone is apparently produced in a tibial gland which presumably evacuates through the pretarsus (Leuthold. 1968).

No ant trail pheromones have yet been characterized structurally but current research in several laboratories is directed toward this objective and it is very probable that the structures of several of these exocrine products will soon be established. In general, the trail

TABLE I

Glandular Sources of Trail Pheromones in the Main Subfamilies in the Formicidae

Subfamily	Source	Species	Reference
Ponerinae	Hind gut	*Termitopone laevigata*	Blum (1966a)
Dorylinae	Hind gut	*Eciton hamatum*	Blum and Portocarrero (1964)
Dorylinae	Hind gut and stomach	*Neivamyrmex* spp.	Watkins (1964)
Myrmicinae	Dufour's gland	*Solenopsis saevissima* (F. Smith)	Wilson (1959)
Myrmicinae	Poison gland	*Tetramorium guineense* (F.)	Blum and Ross (1965)
Myrimicinae	Metathoracic legs	*Crematogaster peringueyi* Emery	Fletcher and Brand (1968)
Dolichoderinae	Pavan's gland	*Iridomyrmex humilis* Mayr	Wilson and Pavan (1959)
Formicinae	Hind gut	*Paratrichina longicornis* (Latr.)	Blum and Wilson (1964)
Formicinae	Hind gut	*Lasius fuliginosus* (Latr.)	Hangartner and Bernstein (1964)

pheromones appear to be relatively stable compounds which should be quite amenable to isolation and structural elucidation. Schneirla and Brown (1950) have demonstrated that in nature *Eciton* trails retain their activities for several weeks, notwithstanding the weathering to which they are subjected. Similarly, Blum *et al.* (1964) have shown that artificial trails prepared from attine poison glands retain their activities for several days at room temperatures. The trail pheromones produced in the poison glands of ants appear to be trace constituents which are not identified with the toxic proteins synthesized by these glandular tissues (Blum and Ross, 1965). The trail substance of *Tetramorium guineense* is quite thermostable and retains its releaser activity in the presence of carbonyl- and carboxylic-derivatizing agents (Blum and Ross, 1965). It has been reported that solutions of the purified trail pheromone isolated from the fire ant *S. saevissima* decompose even when stored at $-20°$ (Walsh *et al.,* 1965). Beroza (1969) has demonstrated, however, that *n*-hexane solutions of the trail pheromone are stable for at least 2 years when stored at 4°C. In all probability, the decomposition of the pheromone observed by Walsh *et al.* (1965) reflects the instability of the solvent methylene chloride, in which the trail substance was dissolved. This chlorinated compound forms acidic hydrolysis products in the presence of water, and if the fire ant trail pheromone is acid-labile, it is probably inactivated fairly rapidly.

Recently, Moser and Silverstein (1967) demonstrated that the trail pheromone of *Atta texana* was composed of at least two substances, one of which was relatively nonvolatile. The volatile fraction could not be detected by the ants after 60 min, whereas the nonvolatile fraction was active for at least 6 days. Presumably, the nonvolatile fraction is oxidatively inactivated since artificial trails were highly active 4 months after preparation, provided that they were held under a vacuum of 0.5 mm of mercury. Significantly, this investigation indicates that ants that lay persistent trails can recruit workers rapidly with a volatile component of their trail pheromone complex while at the same time laying a trail which will be of relatively long duration, especially if it is reinforced with additional trail pheromone. This contrasts markedly with those ephemeral trails which are laid by *Myrmica* workers only during the initial phase of active recruiting (Ayre, 1969).

B. TERMITE TRAIL PHEROMONES

The basis of trail laying in termites has been elucidated by Stuart (1967). In both primitive and higher termites, odor trails are laid from a sternal gland in the ventral part of the fifth abdominal segment (Stuart, 1961). In the primitive genus *Zootermopsis*, trails are laid to breaks in the nest structure and workers are thus recruited rapidly to a specific point in the nest. Alarm behavior is also exhibited by *Zooter-mopsis* workers when they have discovered a nest break, and the subsequent laying of an odor trail is part of a defensive reaction which results in building behavior (Stuart, 1967). The number of workers recruited to the break is related to the intensity of the stimulus, the odor trail. In higher termites, food sources produce low-level alarm reactions when they are discovered and trails are laid from the food finds back to the nest. Thus, in both lower and higher termites, alarm stimuli release trail-laying behavior.

The odor trail pheromone employed by *Reticulitermes virginicus* has been identified by Matsumura *et al.* (1968). The pheromone, *n-cis*-3,*cis*-6,*trans*-8-dodecatriene-l-ol, was isolated from both the termites and the fungus-infected wood on which they feed. The results of this investigation indicated that the decayed wood was a much better source of the pheromone than the termite itself. If, indeed, the termites obtain the trail pheromone from their food source, then this dodecatrienol must be channeled to the sternal gland in order for it to be secreted on an odor trail. Since it is generally believed that odor trail pheromones are biosynthesized in specific exocrine tissues, the occurrence of large amounts of the trail substance of *R. virginicus* in the food source raises some provocative questions, especially in regard to the origins of trail pheromones in termites. Hopefully, subsequent investigations will illuminate the exact role that decaying wood plays in the social biology of *R. virginicus* and other termites.

C. BEE TRAIL PHEROMONES

In an elaborate series of experiments, Lindauer and Kerr (1958) demonstrated that workers of many species of stingless bees in the tribe Meliponini lay trails from food finds to their nests. The trail is laid as a series of droplets which are placed by the scout bee at specific intervals on either vegetation or the ground as the bee moves from the food source to the nest. The distance between the scent marks varies with each species of bee, but in all cases so far examined, the marking pheromones originate in mandibular gland secretions (Lindauer and Kerr, 1960). Recruited bees orient to an "aerial" trail which results from the volatilization of the deposited mandibular gland droplets.

These trails are rapidly reinforced by returning recruits and thus an effective concentration of trail pheromone is maintained during the critical period, the initial recruitment phase. The efficiency of recruiting in the higher Meliponini is as great as that achieved by the honeybee *Apis mellifera.* In addition, stingless bee trails can be laid in the vertical component, whereas the honeybee is able to communicate only horizontal distances (Lindauer and Kerr, 1958). Therefore, stingless bees can recruit workers to flower concentrations in tall tropical trees, but honeybees must necessarily restrict their recruiting activities to the flora near the ground.

The mandibular gland complexes of two species of stingless bees, *Trigona postica* and *T. tubiba*, have recently been analyzed (Blum *et al.,* 1969d). The chief components in the mandibular glands of these species are listed in Table II.

TABLE II

Chemical Composition of the Mandibular Gland
Secretions of *Trigona postica* and *T. tubiba*[a]

Compound	Species	
	T. postica	*T. tubiba*
2-Heptanone	+	+
2-Nonanone	+	+
2-Hendecanone	+	−
2-Tridecanone	+	−
2-Pentadecanone	+	−
2-Heptadecanone	+	−
2-Nonadecanone	+	−
2-Heneicosanone	+	−
2-Tricosanone	+	−
2-Pentacosanone	+	−
Benzaldehyde	+	+
n-Hendecane	+	+
n-Tridecane	+	+

[a]From Blum *et al.* (1969d).

The extensive series of methyl ketones present in the mandibular gland secretion of *T. postica* presents this species with a fingerprint of natural products that so far easily distinguishes it from any other social insect. Indeed, with the exception of 2-heptanone, which is present in the mandibular gland secretion of the honeybee (Shearer and Boch,

1965), none of these alkanones have been previously isolated from bees. Both 2-heptanone (Blum *et al.*, 1963) and 2-tridecanone (Quilico *et al.*, 1957) have been isolated from several species of ants, and benzaldehyde recently has been identified in the mandibular gland secretion of an ant (Blum *et al.*, 1969c). Surprisingly, no terpenoid constituents were detected in either of these species of *Trigona*, although this class of compounds appears to be utilized extensively by the honeybee for communicating.

Both *T. postica* and *T. tubiba* are members of the subgenus *Scaptotrigona* and it is perhaps not surprising that their glandular secretions share several common constituents. Since scent trails are laid with mandibular gland secretions by both species, obviously the trail "language" of *T. postica* differs considerably from that of *T. tubiba*. It has been determined that both 2-heptanone and 2-nonanone are attractants and, in high concentrations, releasers of attack behavior for both species of *Trigona*. On the other hand, benzaldehyde is highly attractive to workers of *T. tubiba*, but does not cause the workers to exhibit the aggressive behavior which characterizes their responses to the two alkanones. Thus, it seems likely that benzaldehyde, which dominates the mandibular gland secretions of *T. tubiba*, functions as the critical volatile element in establishing the aerial trails employed by this species. Similarly, benzaldehyde is attractive to workers of *T. postica* but the potency of this compound as an attractant for this species is markedly enhanced by the presence of the C_{11}–C_{13} ketones (Blum *et al.*, 1969d). Therefore, it appears possible that these two related species of *Trigona* are insulated from following each other's trails (assuming that they are sympatric) because of the specificities of their chemical trails. Since species in the subgenus *Scaptotrigona* lay scent marks about 1–2 m apart, any river which is wider than several meters constitutes a geographical barrier which effectively isolates populations (Kerr, 1960). The development of new species of *Scaptotrigona* appears to have resulted in populations which are isolated by rivers (Kerr, 1960) and it is not unlikely that the chemical compositions of the mandibular gland secretions may have diverged so as to ensure that each population will have its own peculiar trail "language." The importance of such a development is emphasized by the demonstration that overlapping trails laid by normally allopatric species can confuse workers of one species which may then follow either trail (Lindauer and Kerr, 1958). Such confusion could have disastrous consequences in terms of reducing the available foraging population of a colony.

Benzaldehyde would appear to be an ideal pheromone for generating temporary trails to floral concentrations. This aldehyde oxidizes rapidly to benzoic acid which is not behaviorally active for these species. As a consequence, unless the original scent trails are reinforced by supplemental mandibular gland secretions, the trail will rapidly lose its olfactory potency, thus ensuring that trails laid to dissipated food sources are no longer attractive. The utilization of this compound as a trail indicator for these species of *Trigona* contrasts strikingly with its role as a defensive secretion in an ant (Blum *et al.*, 1969c) and in millipedes (Eisner *et al.*, 1963). If social insects have a rather restricted repertoire of natural products with which to regulate their behavior, simple defensive compounds employed by nonsocial insects may be found to be employed routinely as exocrine products by social arthropods. Other examples of common arthropod defensive compounds which function as pheromones in social insects are already known (Bevan *et al.*, 1961). Since benzaldehyde is also utilized by male noctuid moths as an aphrodisiac (Aplin and Birch, 1968), it is evident that even some sexual pheromones are identical to defensive compounds employed by arthropods.

III. ALARM PHEROMONES

Many species of social insects can rapidly communicate colonial disturbances by secreting pheromones from exocrine glands. Maschwitz (1964) has demonstrated that all major groups of hymenopterous insects employ alarm pheromones and Moore (1968) determined that these excitatory substances are secreted by certain species of termites. The widespread distribution of alarm pheromones emphasizes the probable importance of these chemical releasers in the biology of social insects. This conclusion is reinforced by the fact that many species of social insects generate alarm signals from two or more glands (Maschwitz, 1964), a development that would seem to ensure that a powerful olfactory signal can be developed whenever a worker is disturbed. Furthermore, whereas insect pheromones generally represent trace glandular products, the alarm compounds are present in such significant quantities that it has even been possible to identify them unequivocally in the bodies of single insects (McGurk *et al.*, 1966). Their distinction as the most concentrated of the known insect pheromones underscores the critical roles that alarm pheromones must play as regulators of social behavior. The functions of alarm pheromones have been recently reviewed (Blum, 1969a).

The main alarm pheromones isolated from social insects are presented in Table III. Although the list of compounds cataloged in this table represents a survey of relatively few of the thousands of species of social insects, it is probable that these alarm pheromones are representative of this class of compounds as a whole. Wilson and Bossert (1963) predicted that these chemical releasers would have molecular weights between 100–200 and would have short "fading" times after they had been secreted. All the known pheromones are represented by small molecules which appear to be ideal odorants for functioning as rapid generators of an alarm signal. Small carbonyl compounds have been stressed as releasers of alarm behavior, probably because this class of compounds is optimum for generating rapid signals without being highly persistent. The utilization of small molecules which are produced in rather large quantities obviously represents less of an energetic demand on the metabolic resources of an insect than would occur if large molecules were similarly employed.

The rapid cataloging of alarm pheromones has provided considerable insight into the chemical code employed in olfactory communication. Half of the compounds listed in Table III are ketones with a restricted range of seven to nine carbon atoms. Four different octanones are utilized as alarm pheromones and three of these, 4-methyl-3-heptanone, 2-methyl-4-heptanone, and 3-octanone, are isomeric C_8 ketones. Similarly, 4-methyl-2-hexanone is an isomeric form of another alarm pheromone, 2-heptanone. The C_6–C_8 alarm pheromones occur in a wide variety of unrelated ants and bees, a further indication of the optimal functionality that these signal molecules must possess in the social milieu in which they are elaborated. Indeed, the random distribution of 2-heptanone in both unrelated ants (Blum *et al.,* 1963; Moser *et al.,* 1968) and bees (Shearer and Boch, 1965; Blum *et al.,* 1969d) presents cogent evidence for regarding this alkanone as a strikingly representative example of a highly efficacious alarm pheromone. 2-Heptanone, which boils at 151°C, exemplifies the majority of alarm pheromones which, although they show little associative tendencies, possess a moderate vapor pressure. Furthermore, virtually all known alarm pheromones possess functional groups which, while they

TABLE III

Primary Alarm Pheromones Employed by Social Insects

Compound	Species	Family	Reference
Citral	*Lestrimelitta limao*	Apidae	Blum (1966b)
Citral	*Acanthomyops claviger*	Formicidae	Ghent (1961)
Citronellal	*Acanthomyops claviger*	Formicidae	Ghent (1961)
2-Heptanone	*Iridomyrmex pruinosus*	Formicidae	Blum *et al.* (1963)
2-Hexenal	*Crematogaster africana*	Formicidae	Bevan *et al.* (1961)
Isoamyl acetate	*Apis mellifera*	Apidae	Boch *et al.* (1962)
Limonene	*Drepanotermes rubriceps*	Termitidae	Moore (1968)
4-Methyl-3-heptanone	*Pogonomyrmex barbatus*	Formicidae	McGurk *et al.* (1966)
2-Methyl-4-heptanone	*Tapinoma nigerrimum*	Formicidae	Trave and Pavan (1956)
6-Methyl-5-hepten-2-one	*Dolichoderus scabridus*	Formicidae	Cavill and Hinterberger (1960)
4-Methyl-2-hexanone	*Dolichoderus clarki*	Formicidae	Cavill and Hinterberger (1962)
2-Nonanone	*Trigona postica*	Apidae	Blum *et al.* (1969d)
3-Octanone	*Crematogaster peringueyi*	Formicidae	Crewe *et al.* (1969)
Terpinolene	*Amitermes herbetensis*	Termitidae	Moore (1968)
2-Tridecanone	*Acanthomyops claviger*	Formicidae	Regnier and Wilson (1968)

reduce the vapor pressure somewhat compared to the nonsubstituted compounds, nevertheless result in the pheromones being potent olfactants. The obtainment of effective olfactory stimulants by social insects appears to have been achieved often by utilizing simple molecules that derive most of their odorant qualities from the presence of an oxygen atom. In view of the ketonic variations, which have already been expressed on a six to eight carbon theme, it seems likely that the biochemically versatile social insects ultimately will be shown to have exploited further the structural possibilities attainable with simple carbonyl compounds. In so doing, these arthropods may continue to be the sources of compounds which are either new natural products or known only from members of plant taxa. The isolation of 4-methyl-3-heptanone (McGurk *et al.*, 1966) and 3-octanone (Crewe *et al.*, 1969) from ants has emphasized vividly the natural products potential of social insects.

IV. VARIED EXPRESSIONS OF PHEROMONAL SOCIALITY

Recognition of both the existence of pheromones and their potential roles as regulators of sociality finally has permitted biologists to illuminate the basis for many heretofore cryptic aspects of insect behavior. Although the precise mechanisms which control many facets of social insect behavior still constitute *terra incognita*, it has become abundantly clear that many expressions of sociality have pheromonal origins. In all probability, a multitude of chemical signals remains to be discovered as an inevitable consequence of examining insect behavior in terms of chemical communication.

The main classes of recognized social insect pheromones have been examined in a relatively recent review (Wilson, 1965) and the present discussion will be limited mainly to an exploration of the new developments in this field. Particular emphasis will be placed on those pheromones which have been identified chemically and thus permit sociality to be examined in terms of a defined chemical "language." At this juncture it seems inappropriate to attempt to categorize the multitude of newly recognized pheromones which clearly constitutes a potpourri of behavioral regulators.

Sex attractants undoubtedly are possessed by numerous social insects, but only that of the honeybee has been characterized chemically. This compound, *trans*-9-keto-2-decenoic acid, is produced in the mandibular glands of the queen and functions to attract drones during the nuptial flight (Gary, 1962). Although insect sex attractants generally are regarded to function as physiological primers, it appears possible that the sex pheromone of the queen honeybee may also function, under certain circumstances, as a sexual releaser for the drone. Frequently drones that were attracted to the sex attractant on filter papers suspended from a balloon collided and ejaculated on this supportive structure (Blum, 1969b). Obviously, the drone, which is attracted to the aerial target in a fully primed sexual state, is capable of achieving the ultimate reproductive releaser response, the ejaculatory reaction, after the stimulative contact afforded by the taut balloon.

Another pheromone produced by the queen honeybee, *trans*-9-hydroxy-2-decenoic acid, also exhibits slight activity as a sex attractant (Butler and Fairey, 1964). Both this compound and 9-keto-2-decenoic acid, which are produced exclusively in the mandibular glands of queen bees, possess other social functions in the honeybee colony. Their additional roles as exocrine regulators will be described in the subsequent section.

The pheromones produced by worker honeybees have been the subject of extensive investigations in the last several years. Workers possess an abdominal structure, the Nassanoff gland, which they expose while fanning their wings, particularly after locating a new food source. The volatile substances emanating from the Nassanoff gland serve to orient recruits to the newly discovered foraging area. Boch and Shearer (1962) have shown that the terpene geraniol is a major constituent present in the secretion of the Nassanoff gland. They reported that this alcohol is attractive to workers. Subsequently, Boch and Shearer (1964) isolated and identified nerolic and geranic acids from this secretion and demonstrated that fortifying geraniol with these two acids produced a mixture that was almost as attractive to workers as the Nassanoff secretion itself. More recently, Shearer and Boch (1966) isolated a mixture of the isomers of citral from Nassanoff secretion which had been stored for several hours at room temperature. These investigators concluded that the aldehydes were produced during the prolonged storage of the collected secretion. Citral had also been detected in the Nassanoff secretion by Weaver *et al.* (1964) and more recently by Butler and Calam (1969). The latter investigators reported that citral was normally present in the secretion and, indeed, was the constituent primarily responsible for the attractiveness of the Nassanoff volatiles to honeybee workers.

Since the *cis*-terpene, nerolic acid, is the major component present in the Nassanoff secretion (Boch and Shearer, 1964), it does not seem likely that this compound could arise by oxidation of the *trans*-alcohol, geraniol. Indeed, Shearer and Boch (1966) have demonstrated that citral is formed in a nitrogen atmosphere when geraniol-rich wipes of the Nassanoff gland are held at room temperature. Possibly the terpenes in the Nassanoff gland are transformed to different oxidative states by enzymatic action. It may well be that the exact roles of the Nassanoff terpenes in the biology of the honeybee will have to be interpreted in terms of the behavioral propensities of the bees at the time that they are exposed to the glandular products. If the threshold for response to a particular pheromone varies among the individuals in a population, possibly because they are participating in different social undertakings at the time of exposure, it is not inconceivable that a variation in behavioral responsiveness will characterize the reactions of the insects under observation. Free (1968) has demonstrated that either geraniol or Nassanoff gland secretions generally inhibited scout bees from exposing their own Nassanoff glands, a finding which indicates further that the threshold for exposing this gland may vary among individuals which are nevertheless exposed to the same apparent stimuli.

Law *et al.* (1965) characterized the volatile compounds present in the mandibular glands of males in the formicine genera *Lasius* and *Acanthomyops*. Ants in these genera produced acyclic terpenes such as citronellol and apparent homologs of this alcohol. Males collected after a nuptial flight contained reduced amounts of these terpenes compared to the quantities in males which had not yet participated in these flights. These authors concluded that the mandibular gland products of the males may function as sex attractants during the nuptial flights of the different species. However, males were collected by Law *et al.* (1965) after the postnuptial flight at lights several hours after the flight rather than during the flight or shortly after mating had occurred. Thus, the terpenoid decrease noted in the nocturnally collected males may simply reflect a dissipation of the mandibular gland products completely unrelated to the mating activities which occur shortly after the inception of the nuptial flight. Indeed, Kannowski and Johnson (1970) have compared the concentrations of terpenes in *Lasius* males before and during flight as well as after they had mated and found no differences in the levels of these compounds. Furthermore, Kannowski (1963) has observed that *Lasius* males are attracted to the females during the nuptial flight, a fact that indicates it is the females rather than the males that produce sex attractants. Kannowski and

Johnson (1970) have documented the role of a female-derived sex attractant in the mating biology of species in the highly evolved formicine genus *Formica*. On the other hand, males of the genus *Camponotus* appear to initiate a flight response in females by liberating their mandibular gland contents on the nest surface (Holldobler and Maschwitz, 1965). Presumably, female flight becomes possible because the male-derived pheromones reduce the flight thresholds of the primed females on the surface of the nest.

Male bumblebees (*Bombus* spp.) employ their mandibular gland secretions in order to mark territorial sites, which may play an important role in their mating biologies. While the scent marks produced by a male of one species do not attract males of other species and, therefore, function as territorial markers, the marks do attract both males and females of the same species, thus bringing the reproductives of both sexes together and increasing the probability that mating will occur. Probably the scent marks also physiologically prime the males and females for subsequent copulatory activity and thus play the role of sexual stimulants. Bergstrom *et al.* (1968) identified 2,3-dihydro-6-*trans*-farnesol as the main component in the mandibular gland secretion of *Bombus terrestris*. The acetate of this alcohol and ethyl laurate were also present. The role that these male-scenting compounds may play as species isolators was emphasized by Calam (1969), who identified the mandibular gland components present in five species of male bumblebees. In *Bombus agrorum, cis*-hexadec-7-en-l-ol was isolated from the males, but was absent in queens and workers. Another alcohol, hexadec-9-en-l-ol, along with hexadecanol and *n*-tricosane, was present in *B. lapidarius*. A series of ethyl esters dominated by ethyl tetradec-9-enoate characterized the mandibular gland secretion of *B. lucorum*. The hydrocarbons *n*-heneicosane, *n*-tricosane, and *n*-pentacosane accompanied the esters. On the other hand, young males of *B. derhamellus* contained primarily alkanes ($C_{18}H_{38}$–$C_{25}H_{52}$) and alkenes ($C_{20}H_{40}$–$C_{25}H_{50}$). The main compounds present were tricosane, tricosene, pentacosane, and pentacosene. A terpenoid alcohol, probably *trans, trans*-farnesol, constituted the main volatile detected in the heads of male *B. pratorum*. Significantly, although *B. terrestus* and *B. lucorum* are so taxonomically similar as to be considered often as subspecies, the occurrence of unrelated natural products in the males of these two bees strongly supports their designations as two distinct species (Calam, 1969).

While living ants undoubtedly employ a multitude of pheromones

as social regulators, it seems extraordinary to recognize that pheromones play a clear role in signaling the death of an individual. Wilson
et al. (1959) demonstrated that extracts of dead *Pogonomyrmex badius*
workers, when applied to filter paper squares, were treated similarly to
dead workers, and were transported rapidly to refuse piles. Synthetic
oleic acid also produced this response. Blum (1969b) fractionated dead
Solenopsis saevissima workers and determined that the releasers of
necrophoric behavior were restricted to the rich free fatty acid fraction.
Myristoleic, palmitoleic, oleic, and linoleic acids were present in this
fraction and it was determined ultimately that all the acids possessed
necrophoric activity. Thus, the high concentration of unsaturates that
accumulates in the free fatty acid fraction of dead ants appears to be
responsible for the generation of an unequivocal death "signal."
Whether the titer of free fatty acids increases because of the autolytic
catabolism which accompanies necrobiosis rather than by the hydrolytic
activities of bacteria cannot be determined at this juncture. Dead
workers of *S. saevissima* are rich in bacteria capable of rapidly
hydrolyzing triglycerides and thus generating a high concentration of
free fatty acids. Indeed, although the death of a worker may be a
physiological *fait accompli*, it is nevertheless true that living workers do
not treat freshly killed individuals in the same manner as they do one
which has been dead for several hours. A freshly frozen worker, when
presented to its sister ants, is a source of great attention and tactile
treatment and is often taken into the interior of the nest rather than the
refuse pile. Eventually, the cadaverous worker is correctly diagnosed as
dead and transported to the area containing dead workers. In all
probability, a freshly killed worker, lacking sufficient concentration of
free fatty acids, is incapable of presenting an olfactory stimulus to
living workers which can demonstrate to them that it is dead. Subsequent necrobiotic reactions generate the volatile acids that provide the
workers with an olfactory cue that initiates a series of responses
resulting in the dead ant being transported and deposited in a refuse
pile.

V. PHEROMONAL SPECIFICITY

Pheromones would appear to constitute ideal species-isolating
substances for social insects. Chemical releasers which are unique to a
species in a particular environment would ensure that communicative

channels which are functions of volatile agents could not be "read" even by closely related sympatric species. Thus, odor trails generated from food finds to the nest would recruit only workers of the trail-laying species, thereby excluding competition from others. Notwithstanding the logical appeal of the pheromonal specificity hypothesis, it would appear that the pheromones employed by social insects often lack specificity even at the generic level. Since virtually all investigations of pheromonal specificity were undertaken under closely controlled laboratory conditions, there is no way of evaluating the multitude of factors which are operative in the field and which may contribute to insulating the communication channels from interpretation by other species.

A. ALARM PHEROMONES

Alarm pheromones appear to be the least specific of the classes of chemical releasers of social behavior. In several instances, however, what at first appeared to be a lack of specificity because the pheromones had not been identified, was later shown to be due to the same pheromone being employed by species in two different genera. Thus, Wilson and Pavan (1959) reported that dolichoderine species in the genera *Tapinoma* and *Liometopum* are equally sensitive to each others' alarm pheromones. Subsequently, it was demonstrated that *Liometopum* (Casnati *et al.*, 1964) produced the same alarm pheromone as *Tapinoma*, 6-methyl-5-hepten-2-one (Trave and Pavan, 1956). Although another dolichoderine, *Iridomyrmex pruinosus*, reacts strongly to ketones closely related to its own alarm pheromone, 2-heptanone (Blum *et al.*, 1963; Blum *et al.*, 1966), this species is relatively insensitive to 4-methyl-2-hexanone, the alarm pheromone utilized by *Dolichoderus clarki* (Cavill and Hinterberger, 1962). Similarly, *Iridomyrmex humilis* is rather insensitive to 2-heptanone, the alarm pheromone liberated by excited workers of another species in this genus, *I. pruinosus* (Blum *et al.*, 1966). When excited, workers of the myrmicine species *Atta texana* liberate two alkanones, 2-heptanone and 4-methyl-3-heptanone, from their mandibular glands (Moser *et al.*, 1968). Whereas 4-methyl-3-heptanone is highly excitatory for this species, 2-heptanone is not.

Thus, although alarm pheromones are far from species-specific, it is nevertheless very clear that ants have a degree of olfactory acuity which enables them, in the presence of their own excitatory odorant, to release alarm behavior at lower threshold values than is possible with related compounds. Vick *et al.* (1969) have demonstrated that two species in the myrmicine genus *Pogonomyrmex* are much more sensitive to their own alarm pheromone, 4-methyl-3-heptanone, than to any member of a series of closely related ketones. *Pogonomyrmex barbatus* was 10,000 times less sensitive to 2-methyl-3-heptanone than to 4-methyl-3-heptanone, a result which demonstrates clearly that the alarm-releasing efficacy of ketones for this species is neither a simple function of molecular weight nor of vapor pressure. Similar results were obtained by Moser *et al.* (1968) who studied an unrelated myrmicine species *Atta texana,* which also utilizes 4-methyl-3-heptanone as an alarm pheromone. More recently, Blum *et al.* (1969b) have studied the alarm-releasing activities of nearly 100 ketones for *Pogonomyrmex badius.* The results of this investigation established that compounds having both a geometry and shape similar to the natural product, 4-methyl-3-heptanone, were the most effective mimics of the alarm pheromone utilized by this species (Table IV). Neither vapor pressure nor molecular weight were critical determinants of the potencies of the synthetic alarm surrogates. Thus, 2-methyl-, 5-methyl-, and 6-methyl-3-heptanone, which have the same molecular weights and virtually the same boiling points as the natural alarm substance, 4-methyl-3-heptanone, are nevertheless only one-half or one-third as active as releasers of alarm behavior. The most active of the synthetic alarm substances, 4-methyl-3-hexanone, 3-heptanone, and 3-methyl-4-heptanone, possess geometries and shapes quite similar to the natural product 4-methyl-3-heptanone. The results of this investigation can be reconciled most conveniently with the stereochemical theory of olfaction which correlates odor qualities with the dimensions and shapes of molecules rather than with the nature of the functional groups (Amoore, 1962, 1964).

The question of the specificity of alarm pheromones ultimately will have to be interpreted in terms of the resolving powers of the olfactory cells of the antennae. It is clear, even now, that a sophisticated decipherment of closely related molecular forms occurs at the antennal receptor sites. Laboratory analyses demonstrate that although ants perceive a wide variety of ketonic compounds (Blum *et al.,* 1969b), only a few of the test compounds are capable, at low concentrations, of transmitting information that can be rapidly decoded and translated into alarm behavior. Potent alarm releasers have been referred to as olfactorily active, but this expression does not illuminate the *modus operandi* of an alarm pheromone at the receptor level in any obvious way.

TABLE IV

Physical Properties and Activity Ratings of Compounds Evaluated
as Alarm Pheromones for *Pogonomyrmex badius*[a]

Compound	Boiling point ($^{\circ}$C; at 760 mm Hg)	Molecular weight	Activity rating[b]
2-Pentanone	103	86	3
3-Methyl-2-butanone	94	86	2
2-Hexanone	128	100	2
4-Methyl-2-pentanone	117	100	3
Cyclohexanone	156	98	1
2-Heptanone	151	114	2
3-Methyl-2-hexanone	137	114	2
4-Methyl-3-hexanone	136	114	7
3-Hexanone	150	114	5
Cycloheptanone	178	112	1
2-Octanone	173	128	1
3-Octanone	169	128	2
2-Methyl-3-heptanone	158	128	2
4-Methyl-3-heptanone[c]	155	128	7
5-Methyl-3-heptanone	158	128	3
6-Methyl-3-heptanone	163	128	2
3-Methyl-4-heptanone	156	128	4

[a] From Blum *et al.* (1969b).
[b] A rating of 1 denotes the least active and 7 the most active response.
[c] The natural alarm substance of *P. badius.*

B. SEX ATTRACTANTS AND QUEEN SUBSTANCE

Among social insects, the only sex attractant that has been identified is that of the queen honeybee *Apis mellifera*. The sex attractant, *trans*-9-keto-2-decenoic acid, has not been reported to attract males of social insects in any other genus. On the other hand, Butler *et al.* (1967) demonstrated that extracts of queens of *Apis cerana* and *Apis florea* were as attractive to drones as those of the honeybee *A. mellifera*. These investigators also obtained persuasive chemical evidence for the presence of 9-keto-2-decenoic acid in both *A. cerana* and *A. florea*. Recently, Shearer *et al.* (1970) isolated *trans*-9-keto-2-decenoic acid from *Apis dorsata,* the remaining species in the genus *Apis*. Since these species of *Apis* are sympatric in certain areas of Asia where *A. mellifera* has been introduced, it would seem highly likely that drones of the various species would be attracted to virgin queens of all four species.

Obviously, no sex attractant specificity would be possible and it would be necessary for additional mechanisms to be operative in order to prevent interspecific matings. Ruttner and Kaissling (1968) have analyzed in detail the primary mechanisms which may be operative in preventing hybridization between *Apis mellifera* and *A. cerana.*

It would not be unexpected if the sex attractants employed by social insects are not completely species-specific. Among lepidopterous insects, numerous examples of nonspecificity are known, even between species in different genera. Since extensive hybridization does not occur normally even among different species which are attracted to the same sex attractant, it is obvious that factors other than the sex attractant reduce the probability of matings occurring between different species.

Kaissling and Renner (1968) have obtained unequivocal electrophysiological evidence for the specificity of queen substance to honeybees. These investigators demonstrated that specific receptor cells on pore plates on the antennae of all three castes of honeybees are sensitive exclusively to *trans*-9-keto-2-decenoic acid. Thus, it is now possible to interpret the specificity of an insect pheromone in terms of a specialized olfactory unit.

Extracts of queens of *Apis cerana, A. mellifera,* and *A. florea* are capable of inhibiting both the development of worker ovaries and the rearing of queens of *A. mellifera* (Butler, 1966). Presumably, the same substance is present in queens of all three species and the lack of queen substance specificity would not be unexpected.

Watkins and Cole (1966) have demonstrated that workers of six species of Nearctic army ants were attracted to the odors of queens of three of the species. Doryline workers, however, appear to be attracted more strongly to the odor of their own queen than to that of unrelated queens of the same species. Queen odors, which may not necessarily be identical to the doryline queen substance, appear to lack absolute specificity among species in the two genera examined.

C. TRAIL PHEROMONES

The identification of the glandular sources of trail pheromones in species in many formicid genera has permitted investigators to determine both the species and generic specificities of these chemical releasers. Surprisingly, the specificities of the trail pheromones appear to be so unpredictable as to nullify the value of any generalizations which can be formulated. Until the chemical structures of these releasers of trail following are established, it would be premature to attempt to interpret critically the results which have been obtained in the specificity studies. Hopefully, this prospect will be realized in the forseeable future.

Wilson (1962a) examined the specificities of the trail pheromones of three species of *Solenopsis*. The pheromone, which is produced in Dufour's gland, is liberated through the sting during the act of trail laying. The pheromone employed by *Solenopsis xyloni* released strong trail following in *S. saevissima* and *S. geminata*. On the other hand, the trail pheromone of *S. geminata* was only slightly active for *S. xyloni* and released no trail following for workers of *S. saevissima*. Workers of *S. geminata* were unresponsive to smears of the Dufour's glands of *S. saevissima*. Thus, the trail pheromone of *S. geminata* appears to be different from that of *S. saevissima,* whereas the releaser of trail following utilized by *S. xyloni* appears to be similar to those utilized by the other two species of fire ants.

A complete lack of specificity was demonstrated for extracts of the trail pheromone glands of species in four genera in the tribe *Attini* (Blum *et al.,* 1964). Workers of species in this myrmicine tribe produce their trail pheromones in the poison gland (Moser and Blum, 1963). Poison gland extracts of workers in the genera *Cyphomyrmex, Trachymyrmex, Acromyrmex,* and *Atta* were highly active in eliciting trail-following responses of species of all genera. Since these genera represent the broad phylogenetic development of the tribe *Attini,* it appears that the trail pheromone developed by the primitive genera (e.g., *Cyphomyrmex*) has not been altered appreciably during the evolution of this tribe. Field observations, however, demonstrated that although the trails of species in the genera *Acromyrmex* and *Atta* overlapped, the two species invariably adhered to their own trails. Thus, supplemental factors must be present on the odor trails in order to ensure that workers in species of one genus do not violate the trail generated by the other species.

The trail pheromones which are biosynthesized in the poison gland are almost certainly trace substances which accompany the proteinaceous constituents characterizing the venoms of most species of stinging ants (Blum and Ross, 1965). Thus, it would not be surprising if the venom of one species contained, in addition to its own trail pheromone, trace constituents which are identical to the trail pheromones utilized by other species of ants. The unexpected lack of specificity of poison gland secretions could then be interpreted in terms of trace releasers of trail following which are common to different venoms. Indeed, if trace constituents which are common to myrmicine venoms are utilized as trail pheromones by species in different genera, then it is likely that species that do not lay chemical trails may nevertheless contain the trail pheromones of other species in their venom. This appears to be precisely the case for the primitive myrmicine species *Daceton armigerum*. Workers of this ant do not lay chemical trails, but their venom contains a potent trail pheromone for species in the attine genera *Trachymyrmex, Acromyrmex,* and *Atta* (Blum and Portocarrero, 1966). However, extracts of the *Daceton* poison gland do not release trail following in workers of the attine genus *Sericomyrmex,* a result which would be consistent with the conclusion that the *Sericomyrmex* trail pheromone is different from that utilized by species in the other attine genera. Indeed, the trail pheromone of *Sericomyrmex* caused no reactions in workers of the other attine genera, thus demonstrating the first case of trail pheromonal specificity among the taxa in the tribe *Attini* (Blum and Portocarrero, 1966).

Watkins *et al.* (1967) studied the specificities of the trail pheromones utilized by six species of doryline ants. Four of five species of *Neivamyrmex* and a species of *Labidus* followed each other's trails. One species, *Neivamyrmex pilosus,* exhibited complete specificity in its ability to respond only to its own trail pheromone. All but one doryline species however, preferred their own odor trails to those of other species, indicating that some degree of olfactory discrimination is present among most of the army ant species examined. *Neivamyrmex* workers were completely unresponsive to trails prepared from nonarmy ant species (Watkins, 1964).

Transposition studies undertaken with the poison gland extracts of four species of *Monomorium* demonstrate that all degrees of specificity are present. *Monomorium minimum* follows artificial trails prepared from its own poison gland and that of *M. pharaonis* (Blum, 1966a). *M. pharaonis,* however, will follow neither the artificial trails of *M. minimum* nor those of *M. floricola.* A fourth species of *Monomorium, M. antarcticum,* would not follow trails prepared from any of its glandular sources and would not follow trails of the three other species. Thus, among the four species in this genus, one follows the trail of another species as well as its own, two species follow only their own trails, and the fourth species could not be demonstrated to follow any artificial trails, including those prepared from its own glandular tissues. It is possible that the venom of *M. pharaonis* contains, in addition to its own trail pheromone, that of *M. minimum*. In this case, workers of *M. minimum* would be expected to follow *M. pharaonis* trails, but not vice versa, since the venom of *M. minimum* would not contain the trail pheromone of *M. pharaonis* (Blum, 1966a). *M. minimum* is also distinguished from the other species of trail-laying *Monomorium* because workers of this myrmicine respond to artificial trails prepared from the poison glands of a nontrail-laying species, *Cardiocondyla nuda*.

The specificity of the scent trails generated by stingless bees has been examined by Kerr *et al.* (1963). *Trigona (Trigona) spinipes* will not follow trails laid by *Trigona (Scaptotrigona) postica*. These two species of stingless bees are not closely related. When workers of two species of bees in the subgenus *Scaptotrigona* were presented with the opportunity to follow each other's trails, a significant demonstration of specificity was apparent. Whereas workers of *Trigona postica* do not follow trails of *T. xanthotricha,* the latter species easily follows the scent trails of *T. postica*. Thus, *T. xanthotricha,* presumably a more recent species, can "read" the trail language of *T. postica,* an older species which cannot decode the trail language of *T. xanthotricha*.

Blum *et al.* (1969d) have analyzed the secretions which constitute the chemical languages employed by *T. xanthotricha* and *T. postica*. The mandibular gland secretion of *T. xanthotricha,* the source of its trail substances, contains aliphatic hydrocarbons, 2-heptanone, and benzaldehyde. The two carbonyl compounds, especially benzaldehyde, are the key components in the trail language of this species. The scent trails of *T. postica* also contain these two compounds, but, in addition, an extensive series of methyl ketones are present. Apparently, *T. postica* is incapable of decoding the chemical information contained in the *T. xanthotricha* trails because these trails lack the ketonic constituents which normally act as olfactory stimulants of high informational content for *T. postica*. On the other hand, *T. xanthotricha* easily "translates" the trail language of *T. postica* which contains the critical information required to elicit trail following. Although the trail secretion of *T. postica* contains many methyl ketones which are not present in the secretion of *T. xanthotricha,* presumably the latter species derives no informational content from them and, as a consequence, reads the *T. postica* message without difficulty. These results demonstrated for the first time that the species specificity of a pheromone may actually reflect the presence of a mixture of compounds which, collectively, are responsible for the uniqueness of the chemical message.

Ultimately, it may well develop that pheromonal specificity can be interpreted in terms of a series of compounds, all of which are necessary for the accurate transfer of chemical information.

VI. PHEROMONAL PARSIMONY:
A KEY ELEMENT IN THE EVOLUTION OF
SOCIALITY

The increase in social efficiency achieved by the more highly evolved arthropods appears to be correlated closely with their successful exploitation of systems of chemical communication. The structures of insect societies have become more complex as additional integrative mechanisms have been developed to regulate the diverse social interactions which occur between individuals in a large cohesive population. Small ant societies appear to function with relatively few pheromonal stimuli, but larger populations appear to be endowed with a multitude of exocrine glands with which to transmit a variety of specific messages. Although many species of social insects are well endowed with pheromone-producing glands (Pavan and Ronchetti, 1955), the total informational content of the exocrine system would appear to be sufficiently finite as to be behaviorally limiting. On the other hand, this communicative limitation can be eliminated if a single pheromone can subserve more than one social function. Critical examinations of several classes of chemical releasers of social behavior have demonstrated that the Hymenoptera and Isoptera have extended considerably their social horizons by means of a well-developed pheromonal parsimony. Therefore, the communicative systems of advanced societies of insects can be examined now in terms of the multiple functions subserved by a single pheromone as a highly successful mechanism of extending the dimensions of sociality.

A. QUEEN SUBSTANCE

The investigations of Butler and his colleagues have illuminated the extraordinarily versatile role that queen substance plays in the biology of the honeybee. Queen substance, *trans*-9-keto-2-decenoic acid, is produced in the mandibular glands of the queen honeybee and is accompanied by an extensive series of ten carbon acids (Callow *et al.*, 1964). Worker honeybees, which are nonreproductive females, do not produce queen substance in their mandibular glands.

9-Keto-2-decenoic acid partially inhibits ovarian development in worker honeybees (Butler and Fairey, 1963). In addition, a second compound, formerly called inhibitory scent, also acts to inhibit the development of worker ovaries, but it is less active than queen substance and does not act synergistically with this compound. Inhibitory scent has been identified recently as *trans*-9-hydroxy-2-decenoic acid (Butler and Callow, 1968). Like queen substance, the hydroxy acid is also synthesized in the mandibular glands of the queen. A mixture of 9-keto-2-decenoic and 9-hydroxy-2-decenoic acids is completely effective in inhibiting queen rearing by queenless workers in cages, but neither compound is highly active by itself (Butler and Callow, 1968). Butler (1960) has shown that in colonies which are preparing to supersede their queen or to accompany her on a swarm, queen rearing appears to result from a diminished production of queen substance by the queen. Queens which were superseded or swarmed with their workers contained about one-fourth as much queen substance as mated laying queens. In light of the fact that queen rearing is now known to be inhibited by a combination of two fatty acids, it seems likely that the diminution in the production of queen substance that occurs in superseded queens actually reflects a decrease in the quantities of both acids.

Barbier and Pain (1960) reported that 9-keto-2-decenoic acid, in combination with other volatile acids produced in the mandibular glands of the queen, was attractive to caged worker bees. More recently, Butler and Simpson (1967) demonstrated that the keto acid would attract swarming bees which had lost their queen. The workers, however, seldom clustered on cages fortified with this acid and those that did were quite restless. By contrast, 9-hydroxy-2-decenoic acid, although not particularly attractive to swarming bees, nevertheless was responsible for producing quiet clusters when workers located a cage containing this acid. Apparently, swarming honeybees are attracted by the keto acid which originates in the queen, but they are behaviorally stabilized by the presence of the hydroxy acid, also produced by the queen. Thus, while one acid orients the restless workers which are participating in a swarming operation to the queen, another volatile acid functions to tranquilize the workers that have settled with the

swarm. The selective value to the bees of this pheromonal tranquiliza-
tion must be especially great, since it ensures that energy dissipation is
reduced during the critical period when the swarming colony is exposed
and particularly susceptible to environmental stresses.

Gary (1962) demonstrated that 9-keto-2-decenoic acid functioned
as a sex attractant for drone honeybees. Butler and Fairey (1964)
reported that this acid was the only compound produced in the
mandibular gland which would attract drones as effectively as a whole
queen. Although 9-hydroxy-2-decenoic acid is slightly attractive to
drones, this compound appears to play no key role as a drone attractant
during the nuptial flight of the queen honeybee. In addition to its role
as a sex pheromone, the keto acid also functions as an aphrodisiac and
stimulates a drone to mount a queen honeybee when her sting-chamber
is open (Butler, 1967).

9-Keto-2-decenoic acid may represent the ultimate example of
pheromonal parsimony among the social insects. This compound
functions as a worker attractant, inhibits queen cell construction and
ovarian development in workers, and is a potent sex attractant and
aphrodisiac for drones. The successful exploitation of a single pheromone for such diverse functions has enabled the honeybee to extend its
social frontiers considerably without requiring the presence of a large
number of signal compounds.

B. TRAIL AND ALARM PHEROMONES

Wilson (1962a,b) has demonstrated that the trail pheromone of
Solenopsis saevissima is utilized for several unrelated functions. This
pheromone is employed as a regulator of mass foraging and colony
emigration and, in conjunction with a cephalic alarm substance, is
secreted when a worker is agitated. Since the alarm pheromone secreted
from the head of a disturbed fire ant does not attract workers, the
simultaneously liberated trail pheromone probably functions to orient
sister workers to the site of the disturbance (Wilson, 1962b).

Alarm pheromones are also utilized as releasers of several different
types of behavior. Wilson (1958) has reported that the alarm pheromone of *Pogonomyrmex badius* functions as an attractant at low
concentrations and a releaser of alarm behavior at higher concentrations. When a laboratory colony of *Tapinoma sessile* was exposed to a
high concentration of one of their alarm pheromones, 6-methyl-
5-hepten-2-one, the queen and workers emigrated to a new nest site

(Wilson and Pavan, 1959). 2-Heptanone, the alarm pheromone utilized by *Conomyrma pyramica*, releases a multitude of behavioral responses in workers of this dolichoderine species (Blum and Warter, 1966). Workers are attracted over considerable distances to the emission source of this pheromone. If alarm pheromones are employed to recruit new workers to food finds, as suggested by Ayre (1968), these compounds may serve another critical function in formicid biology.

High concentrations of 2-heptanone release digging behavior in excited workers of *C. pyramica*, a reaction also observed with workers of *Pogonomyrmex badius* (Wilson, 1958) which are exposed to their alarm pheromone. Workers of *C. pyramica* and *Acanthomyops claviger* exhibit a peculiar "lurching" or "tremoring" behavior in response to their chemical releasers of alarm behavior (Blum and Warter, 1966; Ghent, 1961). When 2-heptanone is placed near the nest opening of a *C. pyramica* colony, the workers will neither leave nor enter the nest. Excited workers of this frenetic dolichoderine species have been observed to dump pupae on the ground, an aberrant behavioral reaction which results from their exposure to high concentrations of their alarm pheromone (Blum and Warter, 1966). Workers, primed by their alarm pheromone, will bite inanimate objects (McGurk *et al.*, 1966) and challenge the sister workers which approach them (Blum, 1969a). It seems likely that workers of many ant species, after detecting their alarm pheromones, possess such a low aggressive threshold that almost any foreign odor will release attack behavior.

The alarm pheromone employed by the stingless bee *Lestrimelitta limao* has been adapted to function as a key element in the robbing behavior of this species. Workers of *L. limao* rob colonies of stingless bees in the genera *Trigona* and *Melipona* (Moure *et al.*, 1958), and their alarm pheromone, citral (Blum, 1966b), subserves several crucial functions during the robbing act. Citral, which is released by the initial robbers, attracts other workers of *L. limao* to the nest which is under attack. As the number of robbing workers increases, the concentration of citral increases considerably in the plundered nest and organized resistance to the attack on the part of the host workers all but disappears. Blum *et al.* (1969a) have demonstrated that citral is highly effective in releasing escape behavior in workers of stingless bee species normally raided by *L. limao*. On the other hand, species of *Trigona* not normally robbed by *L. limao* appear to maintain their colonial cohesiveness when exposed to citral. Thus, citral functions as a potent alarm pheromone not only for workers of *L. limao,* but also for the stingless bees whose nest is being plundered. Citral appears to be the key to successful robbing by *L. limao,* since it totally disorganizes the host workers in the raided nest, probably by virtue of its ability to act as a powerful olfactory stimulant. Citral, which has been demonstrated to be an effective defensive secretion of ants (Ghent, 1961), probably is also employed by *L. limao* workers in a similar capacity. This compound would function as an excellent topical irritant when applied to the cuticle of an alien bee by a worker of *L. limao.*

C. NASSANOFF GLAND PHEROMONE

The high degree of sociality that has been achieved by the honeybee appears to be well correlated with the extensive development of a versatile chemical communication system. The remarkable parsimony achieved by the queen with her mandibular gland pheromones has been described, but it is evident that the workers also have exploited sociality by employing a single glandular exudate to subserve multiple functions. The Nassanoff gland, an abdominal structure which appears to be more highly developed in the honeybee than in any other social bee (Jacobs, 1925), is now known to be utilized as a scenting organ in a number of vital contexts. Indeed, it is tempting to speculate that the social precision that characterizes the organization of a honeybee colony could never have been achieved without the advantages gained by the ability of the workers to utilize the Nassanoff secretion so parsimoniously.

Renner (1960) has described some of the stimuli which are associated with the eversion of the Nassanoff gland. The Nassanoff secretion is attractive to both workers and the queen. Good forage sources are marked with the Nassanoff exudate, in order to increase the number of recruits to the new food find. Workers frequently scent near the hive entrance, a reaction which helps inexperienced bees to locate their hive (Renner, 1960). Workers which had been lost, scent after returning to the hive and if the location of a hive is changed, workers will begin to scent and fan after they have located their hive. Scenting occurs if the workers are separated from their queen, a situation which normally develops when the queen undertakes a nuptial flight (Renner, 1960). Obviously, the Nassanoff volatiles play an important role in attracting both the workers and the queen under many different circumstances.

Lindauer (1951) reported that swarming scout bees, which select potential nesting sites, scent vigorously at the new sites and attract other workers. Boch (1969) has demonstrated that the formation of the interim swarm can be induced by the Nassanoff volatiles which attract large numbers of workers as well as the queen. Although queen-derived pheromones stabilize the interim swarm, the nomadic cluster of bees owes its origin to the terpenes which are liberated from the Nassanoff glands of the scout bees that have selected the temporary nesting site.

Specialized receptor cells which are highly responsive to the Nassanoff volatiles are present in abundance on the antennae of all three castes of honeybees (Kaissling and Renner, 1968). Thus, the honeybee is capable of selectively processing the informational content of the Nassanoff terpenes, a crucial social development which guaranteed the functionality of the Nassanoff gland, a social organ that arose *de novo* in this insect.

SUMMARY

The primary integrating forces in social insect populations are pheromones. The exploitation of sociality by insects appears to be correlated with the development of a sophisticated system of chemical communications which is capable of regulating a multitude of behavioral responses. The deciphering of the pheromonal code has permitted insect sociality to be examined as the product of specific exocrine compounds whose informational content can be rapidly decoded by target insects. Similar pheromonal classes are utilized by isopterous and hymenopterous species in order to achieve equivalent functions in the social milieu. Sex attractants, trail pheromones, and alarm pheromones, are produced by a multitude of social insects, but the specificities of these exocrine products are highly variable even at the generic level. Because of pheromonal parsimony, the ability of single pheromones to subserve multiple functions in different environmental contexts, social insects have been able to expand their communications without requiring a large chemical language. Ultimately, chemical sociality may be shown to be largely a product of frugality in the use of natural products combined with a well-developed behavioral plasticity.

REFERENCES

Amoore, J. E. (1962). The stereochemical theory of olfaction. 1. Identification of the seven primary odors. *Proc. Sci. Sect. Toilet Goods Assoc., Suppl.* **37**, 1.

Amoore, J. E. (1964). Current status of the steric theory of odor. *Ann. N. Y. Acad. Sci.* **116**, 457.

Aplin, R. T., and Birch, M. C. (1968). Pheromones from the abdominal brushes of male noctuid Lepidoptera. *Nature* **217**, 1167.

Ayre, G. L. (1968). Comparative studies on the behavior of three species of ants (Hymenoptera: Formicidae). I. Prey finding, capture, and transport. *Can. Entomologist* **100**, 165.

Ayre, G. L. (1969). Comparative studies on the behavior of three species of ants (Hymenoptera: Formicidae). II. Trail formation and group foraging. *Can. Entomologist* **101**, 118.

Barbier, M., and Pain, J. (1960). Etude de la sécrétion des glandes mandibulaires des reines et des ouvrières d'abeilles (*Apis mellifera* L.) par chromatographie en phase gazeuse. *Compt. Rend. Acad. Sci.* **250**, 3740.

Bergstrom, G., Kullenberg, B., Ställberg-Stenhagen, S., and Stenhagen, E. (1968). Studies on natural odoriferous compounds. II. Identification of a 2,3-dihydrofarnesol as the main component of the marking perfume of male bumble bees of the species *Bombus terrestris* L. *Arkiv Kemi* **28**, 453.

Beroza, M. (1969). Personal communication.

Bevan, C. W. L., Birch, A. J., and Caswell, H. (1961). An insect repellent from black cocktail ants. *J. Chem. Soc.* p. 488.

Blum, M. S. (1966a). The source and specificity of trail pheromones in *Termitopone, Monomorium* and *Huberia,* and their relation to those of some other ants. *Proc. Roy. Entomol. Soc. (London)* A **41**, 155.

Blum, M. S. (1966b). Chemical releasers of social behaviour. VIII. Citral in the mandibular gland secretion of *Lestrimelitta limao. Ann. Entomol. Soc. Am.* **59**, 962.

Blum, M. S. (1969a). Alarm pheromones. *Ann. Rev. Entomol.* **14**, 57.

Blum, M. S. (1969b). Unpublished data.

Blum, M. S., and Portocarrero, C. A. (1964). Chemical releasers of social behavior. IV. The hind gut as the source of the odor trail pheromone in the Neotropical army ant genus *Eciton. Ann. Entomol. Soc. Am.* **57**, 793.

Blum, M. S., and Portocarrero, C. A. (1966). Chemical releasers of social behavior. X. An attine trail substance in the venom of a non-trail laying myrmicine, *Daceton armigerum* (Latreille). *Psyche* **73**, 150.

Blum, M. S., and Ross, G. N. (1965). Chemical releasers of social behaviour. V. Source, specificity and properties of the odour trail pheromone of *Tetramorium guineense* (F.) (Formicidae: Myrmicinae). *J. Insect Physiol.* **11**, 857.

Blum, M. S., and Warter, S. L. (1966). Chemical releasers of social behavior. VII. The isolation of 2-heptanone from *Conomyrma pyramica* and its *modus operandi* as a releaser of alarm and digging behavior. *Ann. Entomol. Soc. Am.* **59**, 774.

Blum, M. S., and Wilson, E. O. (1964). The anatomical source of the trail pheromones in formicine ants. *Psyche* **71**, 28.

Blum, M. S., Warter, S. L., Monroe, R. S., and Chidester, J. C. (1963). Chemical releasers of social behaviour. I. Methyl-*n*-amyl ketone in *Iridomyrmex pruinosus* (Roger). *J. Insect Physiol.* **9**, 881.

Blum, M. S., Moser, J. C., and Cordero, A. D. (1964). Chemical releasers of social behavior. II. Source and specificity of the odor-trail substance in four attine genera (Hymenoptera: Formicidae). *Psyche* **71**, 1.

Blum, M. S., Warter, S. L., and Traynham, J. G. (1966). Chemical releasers of social behaviour. VI. The relation of structure to activity of ketones as releasers of alarm behaviour for *Iridomyrmex pruinosus* (Roger). *J. Insect Physiol.* **12**, 419.

Blum, M. S., Crewe, R. M., Kerr, W. E., Keith, L. H., Garrison, A. W., and Walker, M. M. (1969a). Unpublished data.

Blum, M. S., Doolittle, R. E., and Beroza, M. (1969b). Unpublished data.

Blum, M. S., Padovani, F., Curley, A., and Hawk, R. E. (1969c). Benzaldehyde: defensive secretion of a harvester ant. *Comp. Biochem. Physiol.* **29**, 461.

Blum, M. S., Padovani, F., Kerr, W. E., Doolittle, R. E., and Beroza, M. (1969d). Unpublished data.

Boch, R. (1969). Personal communication.

Boch, R., and Shearer, D. A. (1962). Identification of geraniol as the active component in the Nassanoff pheromone of the honeybee. *Nature* **194**, 704.

Boch, R., and Shearer, D. A. (1964). Identification of nerolic and geranic acids in the Nassanoff pheromone of the honeybee. *Nature* **202**, 320.

Boch, R., Shearer, D. A., and Stone, B. C. (1962). Identification of iso-amyl acetate as active component in the sting pheromone of the honeybee. *Nature* **195**, 1018.

Butler, C. G. (1960). The significance of queen substance in swarming and supersedure in honeybee (*Apis mellifera* L.) colonies. *Proc. Roy. Entomol. Soc. (London)* A **35**, 129.

Butler, C. G. (1966). Die Wirkung von Königinnen Extrakten verschiendener sozialer Insekten auf die Aufzucht von Königinnen und de Entwicklung der Ovarien von Arbeiterinnen der Honigbiene (*Apis mellifera* L.). *Z. Bienenforsch.* **8**, 143.

Butler, C. G. (1967). A sex attractant acting as an aphrodisiac in the honeybee (*Apis mellifera* L.). *Proc. Roy. Entomol. Soc. (London)* A **42**, 71.

Butler, C.G., and Calam, D. H. (1969). Pheromones of the honeybee- The secretion of the Nassanoff gland of the worker. *J. Insect Physiol.* **15**, 237.

Butler, C. G., and Callow, R. K. (1968). Pheromones of the honeybee (*Apis mellifera* L.): the "inhibiting scent" of the queen. *Proc. Roy. Entomol. Soc. (London)* A **43**, 62.

Butler, C. G., and Fairey, E. M. (1963). The role of the queen in preventing oogenesis in worker honeybees. *J. Apicult. Res.* **2**, 14.

Butler, C. G., and Fairey, E. M. (1964). Pheromones of the honeybee: Biological studies of the mandibular gland secretion of the queen. *J. Apicult. Res.* **3**, 65.

Butler, C. G., and Simpson, J. (1967). Pheromones of the queen honeybee (*Apis mellifera* L.) which enable her workers to follow her when swarming. *Proc. Roy. Entomol. Soc. (London)* A **42**, 149.

Butler, C. G., Calam, D. H., and Callow, R. K. (1967). Attraction of *Apis mellifera* drones by the odours of the queens of two other species of honeybees. *Nature* **213**, 423.

Calam, D. H. (1969). Species and sex-specific compounds from the heads of male bumblebees (*Bombus* spp.). *Nature* **221**, 856.

Callow, R. K., Chapman, J. R., and Paton, P. N. (1964). Pheromones of the honeybee: chemical studies of the mandibular gland secretion of the queen. *J. Apicult. Res.* **3**, 77.

Casnati, G., Pavan, M., and Ricca, A. (1964). Richerche sul secreto delle glandole anali di *Liometopum microcephalum* Panz. *Boll. Soc. Entomol. Ital.* **94**, 147.

Cavill, G. W. K., and Hinterberger, H. (1960). The chemistry of ants. IV. Terpenoid constituents of some *Dolichoderus* and *Iridomyrmex* species. *Australian J. Chem.* **13**, 514.

Cavill, G. W. K., and Hinterberger, H. (1962). Dolichoderine ant extractives. *Proc. Intern. Congr. Entomol, 11th, Vienna* **3**, 53.

Crewe, R. M., Brand, J. M., and Fletcher, D. J. C. (1969). Identification of an alarm pheromone in the ant *Crematogaster peringueyi* Emery. *Ann. Entomol. Soc. Am.* **62**, 1212.

Eisner, H. E., Eisner, T., and Hurst, J. J. (1963). Hydrogen cyanide and benzaldeyde produced by millipedes. *Chem. Ind. (London)* p. 124.

Fletcher, D. J. C., and Brand, J. M. (1968). Source of the trail pheromone and method of trail laying in the ant *Crematogaster peringueyi. J. Insect Physiol.* **14**, 783.

Free, J. B. (1968). The conditions under which foraging honeybees expose their Nasanov gland. *J. Apicult. Res.* **7**, 139.

Gabba, A. (1967). Aspetti dell' organizzazione degli insetti sociali. 2. La sostanza della traccia nei Formicidae. *Natura (Milan)* **58**, 150.

Gary, N. E. (1962). Chemical mating attractants in the queen honeybee. *Science* **136**, 773.

Ghent, R. L. (1961). Adaptive refinements in the chemical defense mechanisms of certain Formicidae. Ph.D. thesis, 88 pp. Cornell Univ., Ithaca, New York.

Hangartner, W., and Bernstein, St. (1964). Uber die Geruchsspur von *Lasius fuliginosus* zwischen Nest und Futterquelle. *Experientia* **20**, 396.

Hölldobler, B. and Maschwitz, U. (1965). Der Hochzeitsschwarm der rossameise *Camponotus herculeanus* L. (Hym. Formicidae). *Z. Vergleich. Physiol.* **50**, 551.

Jacobs, W. L. (1925). Das Duftorgan von *Apis mellifica* und ahnliche Hautdrusen sozialer und solitarer Apiden. *Z. Morphol. Oekol. Tiere* **3**, 1.

Kaissling, K.-E., and Renner, M. (1968). Antennale Rezeptoren fur Queen Substance und Sterzelduft bei der Honigbiene. *Z. Vergleich. Physiol.* **59**, 357.

Kannowski, P. B. (1963). The flight activities of formicine ants. *Symp. Genet. Biol. Ital.* **12**, 74.

Kannowski, P. B., and Johnson, R. L. (1970). Male patrolling behaviour and sex attraction in ants of the genus *Formica. Animal Behavior* **17**, 425.

Kerr, W. E. (1960). Evolution of communication in bees and its role in speciation. *Evolution* **14**, 386.

Kerr, W. E., Ferreira, A., and de Mattos, N. S. (1963). Communication among stingless bees: Additional data (Hymenoptera: Apidae). *J. N. Y. Entomol. Soc.* **71**, 80.

Law, J. H., Wilson, E. O., and McCloskey, J. A. (1965). Biochemical polymorphism in ants. *Science* **149**, 544.

Leuthold, R. H. (1968). A tibial gland scent trail and trail-laying behavior in the ant *Crematogaster ashmeadi* Mayr. *Psyche* **75**, 233.

Lindauer, M. (1951). Bientanzen in der Schwarmtraube. *Naturwissenschaften* **22**, 509.

Lindauer, M., and Kerr, W. E. (1958). Die gegenseitige Verstandingung bei den stachellosen Bienen. *Z. Vergleich. Physiol.* **41**, 405.

Lindauer, M., and Kerr, W. E. (1960). Communication between the workers of stingless bees. *Bee World* **41**, 29.

McGurk, D. J., Frost, J., Eisenbraun, E. J., Vick, K., Drew, W. A., and Young, J. (1966). Volatile compounds in ants: identification of 4-methyl-3-heptanone from *Pogonomyrmex* ants. *J. Insect Physiol.* **12**, 1435.

Maschwitz, U. W. (1964). Gefahrenalarmestoffe und Gefahrenalarmierung bei sozialen Hymenoptera. *Z. Vergleich. Physiol.* **47**, 596.

Matsumura, F., Coppel, H. C., and Tai, A. (1968). Isolation and identification of termite trail-following pheromone. *Nature* **219**, 963.

Moore, B. P. (1968). Studies on the chemical composition and function of the cephalic gland secretion in Australian termites. *J. Insect Physiol.* **14**, 33.

Moser, J. C., and Blum, M. S. (1963). Source and potency of the trail marking substance of the Texas leaf-cutting ant. *Science* **140**, 1228.

Moser, J. C., and Silverstein, R. M. (1967). Volatility of the trail marking substance of the town ant. *Nature* **215**, 206.

Moser, J. C., Brownlee, R. G., and Silverstein, R. M. (1968). The alarm pheromones of *Atta texana. J. Insect Physiol.* **14**, 529.

Moure, J. S., Nogueira-Neto, P., and Kerr, W. E. (1958). Evolutionary problems among Meliponinae. *Proc. Intern. Congr. Entomol., 10th Montreal* **2**, 481.

Pavan, M. (1955). Studi sui Formicidae. I. Contributo all conoscenza degli organi gastrali dei *Dolichoderinae. Natura (Milan)* **46**, 133.

Pavan, M., and Ronchetti, G. (1955). Studi sulla morfologia esterna e anatomia interna dell'operaia di *Iridomyrmex humilis* Mayr e ricerche chimiche e biologiche sulla iridomirmecina. *Atti Soc. Ital. Sci. Nat. Museo Civico Storia Nat. Milano* **94**, 379.

Quilico, A., Piozzi, F., and Pavan, M. (1957). Ricerche chimiche sui Formicidae; sostanze prodotte dal *Lasius (Chthonolasius) umbratus* Nyl. *Rev. Cienc. Ist. Lombardo* **91**, 271.

Regnier, F. E., and Wilson, E. O. (1968). The alarm-defense system in the ant *Acanthomyops claviger. J. Insect Physiol.* **14**, 955.

Renner, M. (1960). Das Duftorgan der Honigbiene und die physiologische Bedeutung ihren Lockstoffes. *Z. Vergleich. Physiol.* **43**, 411.

Ruttner, F., and Kaissling, K. -E. (1968). Uber die interspecifische Wirkung des Sexuallstoffes von *Apis mellifica* und *Apis cerana. Z. Vergleich. Physiol.* **59**, 362.

Schneirla, T. C., and Brown, R. Z. (1950). Army ant life and behavior under dry season conditions. 4. Further investigation of cyclic processes in behavioral and reproductive functions. *Bull. Am. Mus. Nat. Hist.* **95**, 263.

Shearer, D. A., and Boch, R. (1965). 2-Heptanone in the mandibular gland secretion of the honeybee. *Nature* **206**, 530.

Shearer, D. A., and Boch, R. (1966). Citral in the Nassanoff pheromone of the honeybee. *J. Insect Physiol.* **12**, 1513.

Shearer, D. A., Boch, R., Morse, R., and Laigo, F. M. (1970). Occurrence of 9-oxo-*trans*-dec-2-enoic acid in queens of *Apis dorsata* and *Apis cerana. J. Insect Physiol.* In press.

Stuart, A. M. (1961). Mechanism of trail-laying in two species of termites. *Nature* **189**, 419.

Stuart, A. M. (1967). Alarm, defense and construction relationships in termites. *Science* **156**, 1123.

Trave, R., and Pavan, M. (1956). Veleni degli insetti. Principi estratti dalla formica *Tapinoma nigerrimum.* Nyl. *Chim. Ind. (Milan)* **38**, 1015.

Vick, K. W., Drew, W. A., Eisenbraun, E. J., and McGurk, D. J. (1969). Comparative effectiveness of aliphatic ketones in eliciting alarm behavior in *Pogonomyrmex barbatus* and *P. comanche. Ann. Entomol. Soc. Am.* **62**, 380.

Walsh, C. T., Law, J. H., and Wilson, E. O. (1965). Purification of the fire ant trail substance. *Nature* **207**, 320.

Watkins, J. F. (1964). Laboratory experiments on the trail following of army ants of the genus *Neivamyrmex* (Formicidae: Dorylinae). *J. Kansas Entomol. Soc.* **37**, 22.

Watkins, J. F., and Cole, T. W. (1966). The attraction of army ant workers to the secretions of their queens. *Texas J. Sci.* **18**, 254.

Watkins, J. F., Cole, T. W., and Baldridge, R. S. (1967). Laboratory studies on interspecies trail following and trail preference of army ants (Dorylinae). *J. Kansas Entomol. Soc.* **40**, 146.

Weaver, N., Weaver, E. C., and Law, J. H. (1964). The attractiveness of citral to foraging honeybees. *Texas Agr. Expt. Sta. Progr. Rept.* **2324**, 1.

Wilson, E. O. (1958). A chemical releaser of alarm and digging behavior in the ant *Pogonomyrmex badius* (Latreille). *Psyche* **65**, 41.

Wilson, E. O. (1959). Source and possible nature of the odor trail of the fire ant *Solenopsis saevissima* (Fr. Smith). *Science* **129**, 643.

Wilson, E. O. (1962a). Chemical communication among workers of the fire ant *Solenopsis saevissima* (Fr. Smith). 1. The organization of mass foraging. *Animal Behav.* **10**, 134.

Wilson, E. O. (1962b). Chemical communication among workers of the fire ant *Solenopsis saevissima* (Fr. Smith). 3. The experimental induction of social responses. *Animal Behav.* **10**, 159.

Wilson, E. O. (1965). Chemical communication in the social insects. *Science* **149**, 1064.

Wilson, E. O., and Bossert, W. H. (1963). Chemical communication among animals. In: *Recent Progr. Hormone Res.* **19**, 673.

Wilson, E. O., and Pavan, M. (1959). Source and specificity of chemical releasers of social behavior in dolichoderine ants. *Psyche* **66**, 70.

Wilson, E. O., Durlach, N., and Roth, L. M. (1959). Chemical releasers of necrophoric behavior in ants. *Psyche* **65**, 108.

ARTHROPOD DEFENSIVE SECRETIONS*

J. Weatherston and
J. E. Percy

I. INTRODUCTION

In a discussion of insect secretions Melander and Brues (1906) refer to two types: "defensive malodorous highly volatile liquids developed principally to repel predacious enemies, and alluring sweet scented or sweet tasting fluids used to attract the two sexes of a species, or the individuals of a community. . .". Much research has been directed in the last two decades to the chemistry and mode of action of such secretions which control insect behavior, and further differentiation of these "chemical messengers" has been estab lished. When the communication is intraspecific the chemical stimulus is known as a pheromone, the term having been coined in 1959 by Karlson and Butenandt. To date the pheromones which have been differentiated are sex pheromones, trail pheromones, swarming phero- mones, alarm pheromones, and territorial marking pheromones; and the literature pertaining to these compounds has been extensively

*Insect Pathology Research Institute, Canadian Department of Fisheries and Forestry, Sault Ste. Marie, Ontario, Canada (Contribution number 133).

95

reviewed in recent years (Karlson and Butenandt, 1959; Wilson and Bossert, 1963; Jacobson, 1965; Wilson, 1965; Cavill and Robertson, 1965; Jacobson, 1966; Butler, 1967; Regnier and Law, 1968; and Blum, 1969).

Of equal importance to these pheromones are those substances which provide chemical communication between different species, although relatively little is known about such interspecific communication. In the realm of the arthropods with its ever present predator–prey relationship the chemical defensive secretion is of the utmost importance. Usually the chemical compound used is of such a type that there is no doubt as to the message it communicates. The literature up to the end of 1966 on the chemical aspects of such defensive secretions has been the topic of several reviews (Roth and Eisner, 1962; Schildknecht et al., 1964; Cavill and Robertson, 1965; Jacobson, 1966; Eisner and Meinwald, 1966; and Weatherston, 1967).

The present review surveys the types of compounds used by arthropods for their defense with discussion of various aspects of their chemistry and biosynthesis. A certain amount of confusion exists in the relationship between "defensive secretions" and "alarm pheromones"; for the present report it is assumed that several of the so-called "alarm pheromones," for example, citral, 4-methyl-3-heptanone, etc. also function as defensive substances. Venoms or other toxins which are administered by stinging or biting are not included.

Arthropods, of all the land-dwelling animals, have the most diverse and probably the best evolved chemical defenses known to man. The chemicals used are usually manufactured in specialized exocrine glands and stored in saclike reservoirs; they may be discharged in gaseous form or as a fine spray, may simply ooze out onto the arthropod's integument as a liquid which volatilizes quickly, or may even be foamy substances. Compounds in this category form the major part of our knowledge of arthropod defensive chemicals, but an equally important area is that of nonexocrine repellent materials found in the blood and other parts of the body.

The phylum Arthropoda is divided into six classes, four of which, namely, Diplopoda (millipedes), Chilopoda (centipedes), Arachnida (spiders, etc.), and Insecta, have been the subject of chemical investigations with regard to defensive secretions. As can be seen from Fig. 1, considering solely the Insecta, of which there are at present 28 orders comprising several million species, the study of arthropod defensive mechanisms can be regarded as being in an early embryonic stage.

It is probable that the first chemical approach to insect defensive

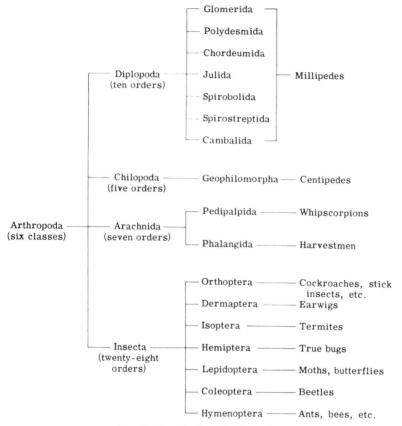

FIG. 1. Classification of arthropods.

secretions dates back to the seventeenth century when Fisher (1670) investigated the distillate of wood ants and came to the conclusion (erroneously) that it was acetic acid. Since then the study of these secretions has expanded rapidly and today over 100 compounds have been isolated and identified. The diversity of the nature of the compounds and the incompleteness of the studies within given groups of arthropods makes classification arbitrary whether it is considered from a chemical or taxonomic standpoint, hence the following are the headings under which the defensive substances are discussed: (*a*) terpenoid and related compounds, (*b*) benzoquinones, (*c*) aromatic compounds, (*d*) miscellaneous compounds including aliphatic hydrocarbons, acids, esters and carbonyl compounds, (*e*) steroids, and (*f*) nonexocrine materials.

II. TERPENOID AND RELATED COMPOUNDS

As can be seen from Fig. 2 the family Formicidae (ants) is divided into seven subfamilies, three of which, namely, the Myrmicinae, the Formicinae, and the Dolichoderinae, have so far been shown to contain genera whose species utilize terpenoid and related compounds as their defensive substances.

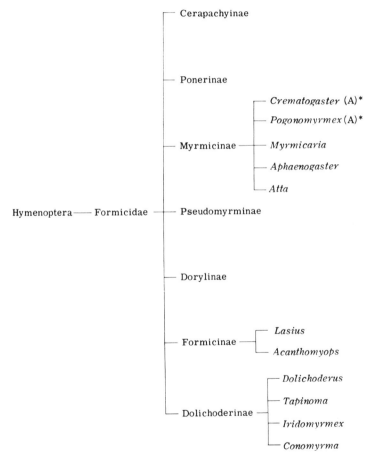

FIG. 2. Genera of ants known to contain terpenoid and related defensive substances. An asterisk indicates that these species do not contain terpenes but only alkanones.

Table I taken in conjunction with Figs. 3–5 shows the species of ants which have been examined and their terpenoid secretions. Figures 3 and 4 show that with the exception of limonene (11) the terpenes isolated from formicine and myrmicine ants are acyclic mono- and sesquiterpenes; whereas Fig. 5 shows that the dolichoderine ants manufacture only cyclopentanoid monoterpenes. Since the last reviews of insect terpenes by Cavill (1960) and Weatherston (1967) several new compounds have been isolated and characterized.

These new studies within the subfamily Formicinae constituted a more thorough reinvestigation of three species, namely, *Acanthomyops claviger, Lasius (Dendrolasius) fuliginosus,* and *L. umbratus.* Blum *et al.* (1968b) have reported that *L. umbratus* also contains in addition to *n*-hendecane and *n*-tridecan-2-one isolated by Quilico *et al.* (1957a), the terpenes citronellal (3.4%) and citronellol (85%). This was the first report of citronellol being an ant defensive substance; however, the same workers (Blum *et al.,* 1968a) later reported that this compound is also found in the defensive secretion of the myrmicine ants *Atta capiguara* and *A. laevigata.*

Work which began over 10 years ago with the isolation by Pavan (1956) of a terpenoid ant constitutent known as dendrolasin, later identified by Quilico *et al.* (1957b) as β-(4,8-dimethylnona-3,7-dienyl)-furan (5) has progressed with the report by Bernardi *et al.* (1967) that the mandibular gland secretion of *L. fuliginosus* contains in addition to dendrolasin the terpenes citral (both isomers) (3), farnesal (7), and a new compound named perillen. Based on the evidence from the mass spectrum of perillen and a careful comparison with the fragmentation pattern of dendrolasin the Italian workers have identified perillen as β-(4-methylpenta-3-enyl) furan (6). The isolation of farnesal (7) is of particular interest since this is only the second report of an acyclic sesquiterpenoid being produced by ants for their defense; the first was in a publication by Cavill *et al.* (1967) noting that α-farnesene (12) is the sole constituent in Dufour's gland (accessory poison gland) of *Aphaenogaster longiceps.*

A very detailed study of the glandular secretion of *Acanthomyops claviger* by Regnier and Wilson (1968) has shown that this ant contains two terpenoidlike compounds, 2,6-dimethyl-5-hepten-l-ol (8) and the corresponding aldehyde (9) in the mandibular gland secretion in addition to citronellal and citral previously isolated by Chadha *et al.* (1962). Although the authors, Regnier and Wilson (1968) are of the opinion that these materials function in the alarm-defense system of the ant, it is also possible that they may be biosynthetic by-products.

TABLE I

Terpenoid and Related Compounds Isolated from Ants[a]

Hymenoptera: Formicidae	Compound	References
Formicinae		
Acanthomyops claviger*	1, 3, 8, 9	Chadha et al. (1962), Regnier and Wilson (1968)
Lasius (Dendrolasius) fuliginosus*	3, 5, 6, 7	Bernardi et al. (1967)
Lasius spathepus	1	Blum (1969)
Lasius umbratus*	1, 2	Quilico et al. (1957a), Blum et al. (1968b)
Myrmicinae		
Aphaenogaster longiceps	12	Cavill et al. (1967)
Atta capiguara*	2	Blum et al. (1968a)
Atta laevigata*	2	Blum et al. (1968a)
Atta sexdens*	3, 10	Blum et al. (1968a)
Atta sexdens rubropilosa	3	Butenandt (1959), Butenandt et al. (1959)
Myrmicaria natalensis	11	Grünanger et al. (1960)
Dolichoderinae		
Conomyrma pyramicus bicolor	3	McGurk et al. (1968)
Conomyrma pyramicus flavopectus*	16	McGurk et al. (1968)
Conomyrma pyramicus pyramicus	3, 16	McGurk et al. (1968)
Dolichoderus clarki	17, 18	Cavill and Hinterberger (1960)
Dolichoderus dentata	17	Cavill and Hinterberger (1960)
Dolichoderus scabridus	3, 14, 16, 17	Cavill and Hinterberger (1960)
Iridomyrmex conifer	4, 16	Cavill et al. (1956)
Iridomyrmex detectus	4, 16	Cavill et al (1956), Cavill and Hinterberger (1960)
Iridomyrmex humilis	13	Cavill et al. (1956)
Iridomyrmex myrmecodiae	17	Cavill and Hinterberger (1960)
Iridomyrmex nitidiceps	4, 16	Cavill and Hinterberger (1960)
Iridomyrmex nitidus	14, 15	Cavill et al. (1956), Cavill and Clark (1967)
Iridomyrmex pruinosus analis	13, 16	McGurk et al. (1968)
Iridomyrmex rufoniger	4, 16, 17	Cavill and Hinterberger (1960)
Tapinoma nigerrimum	4, 16, 19	Trave and Pavan (1956), Pavan and Trave (1958)
Tapinoma sessile	14, 16	McGurk et al. (1968)

[a] An asterisk indicates that secretion contains other components.

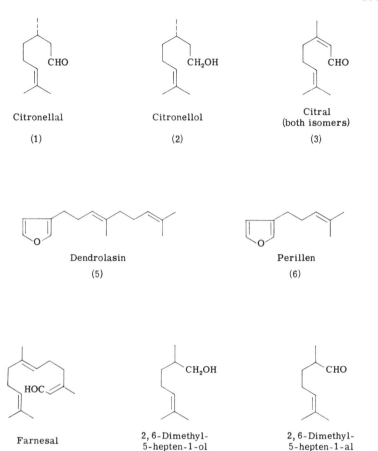

FIG. 3. Terpenoids and related compounds isolated from ants of the subfamily Formicinae. (For structure **4**, see Fig. 5).

Turning now to the dolichoderine ants, two recent publications have added another iridolactone to the list of cyclopentanoid monoterpenes used by these ants; and the stereochemical variation of iridodial (**16**, see Fig. 5) within the subfamily has been discussed. As reported by McGurk *et al.* (1968) there are four isomers of iridodial (**16a–d**) because of epimerization at the carbon atoms alpha to both carbonyl groups.

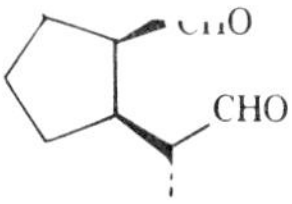

(16a)	(16b)	(16c)	(16d)

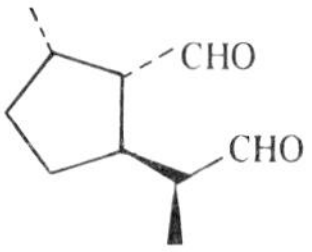

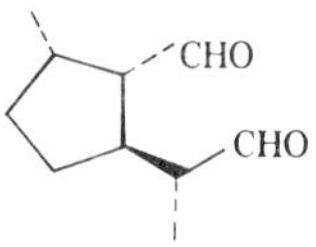

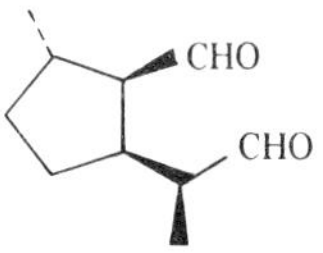

Investigating three dolichoderine species, *Iridomyrmex pruinosus analis*, *Tapinoma sessile*, and *Conomyrma pyramicus pyramicus* these authors found that each species contained more than one isomer of iridodial. Each of the species contained more than 80% of one isomer and this was different for each species examined. *I. pruinosus analis* contained more than 80% δ-iridodial **(16d)**, whereas the major component of *T. sessile* and *C. pyramicus pyramicus* was α-iridodial **(16a)** and β-irido-dial **(16b)**, respectively. The significance of the synthesis of different isomers by members of this subfamily is not as yet understood.

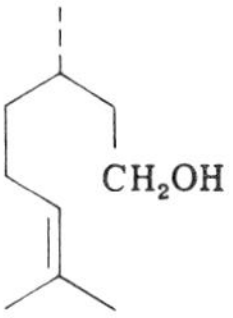

Citronellol

(2)

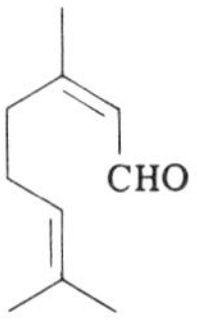

Citral
(both isomers)

(3)

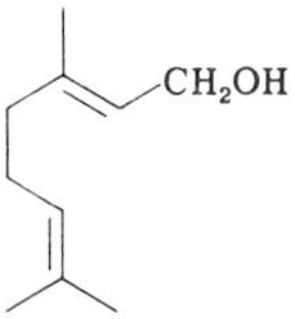

Geraniol

(10)

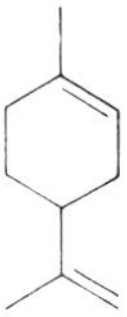

Limonene

(11)

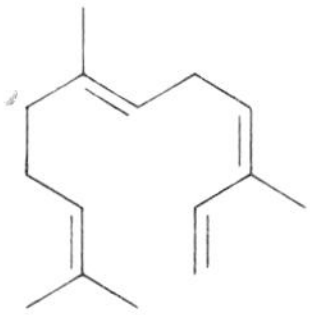

α-Farnesene

(12)

Fig. 4. Terpenoids and related compounds isolated from ants of the subfamily Myrmicinae.

Cavill and Clark (1967) have recently reported that the anal gland secretion of *I. nitidus* contains isodihydronepetolactone **(15)**, in addition to isoiridomyrmecin **(14)**.

In both the above reported studies the structural and stereochemical elucidations were carried out by correlation with the known series of nepetalinic acids.

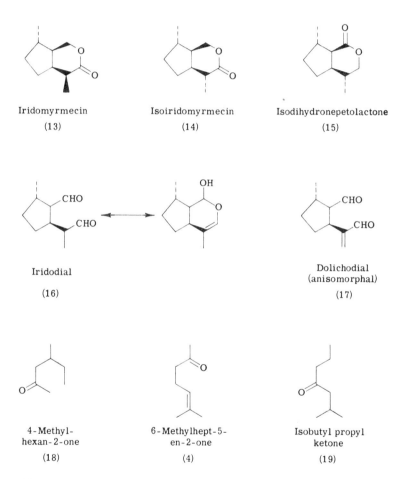

Iridomyrmecin (13)

Isoiridomyrmecin (14)

Isodihydronepetolactone (15)

Iridodial (16)

Dolichodial (anisomorphal) (17)

4-Methyl-hexan-2-one (18)

6-Methylhept-5-en-2-one (4)

Isobutyl propyl ketone (19)

FIG. 5. Terpenoids and related compounds isolated from ants of the subfamily Dolichoderinae.

BIOSYNTHESIS OF TERPENOID
DEFENSIVE SUBSTANCES

Experimental verification of postulated biosynthetic routes leading to terpenoid defensive substances has not been forthcoming. Tracer studies by Happ and Meinwald (1965, 1966) have shown that the alicyclic monoterpenes citral (3) and citronellal (1) isolated from *Acanthomyops claviger* incorporated 1-^{14}C-acetate, 2-^{14}C-acetate, and 2-^{14}C-mevalonate, but not 1-^{14}C-mevalonate suggesting that the normal pathway of terpene biogenesis is being followed (Richards and Hendrickson, 1964). The nonincorporation of 1-^{14}C-mevalonate occurs because activity is lost during decarboxylation in the formation of Δ^3-isopentenylpyrophosphate in the early stage of terpene formation.

The sesquiterpenoid dendrolasin (5) is most probably derived by the normal pathway via farnesol or farnesal (7) followed by ring closure to form the furan. This hypothesis has received some indirect verification with the isolation (Bernardi *et al.*, 1967) of farnesal (7), occurring with dendrolasin in the mandibular glands of *Lasius fuliginosus*. The occurrence of perillen (6) and citral (3) in the same gland raises the question of whether perillen is a by-product of dendrolasin biosynthesis or an intermediate in dendrolasin biogenesis.

Several routes have been proposed for the biosynthesis of the cyclopentanoid monoterpenes and the aliphatic ketones which occur with them in dolichoderine ants; the most feasible is that of Cavill (1960) and Cavill and Robertson (1965) which is shown schematically in Fig. 6. The methylheptanone (4) could arise from a reverse aldol-type reaction on citral (3). In the main scheme citral would be enzymatically reduced to citronellal (1), terminal oxidation of which would give 2,6-dimethyloct-2-ene-1,8-dial, which on cyclization would yield iridodial (16). This compound occupies a key position in the scheme, for easy transformations would lead to dolichodial (17) and the iridolactones (13–15). Evidence supporting Cavill's hypothesis has come from a recent report (McGurk *et al.*, 1968) that for the first time iridodial (16) has been found to co-occur with an iridolactone in an ant. In *Iridomyrmex pruinosus analis* iridomyrmecin (13) is present, while *Tapinoma sessile* contains isoiridomyrmecin (14); that is, in each case the iridolactone possesses the same stereochemistry as the predominant iridodial isomer in that ant.

Fɪɢ. 6. Postulated route to the biosynthesis of arthropod cyclopentanoid monoterpenes.

III. ARTHROPOD BENZOQUINONES

Benzoquinones have been the most widely found components of arthropod defensive substances, having been isolated from over fifty species of at least seven orders belonging to the classes Arachnida, Diplopoda, and Insecta. The benzoquinones which have been identified so far as defensive agents are shown in Fig. 7 and their occurrence has been reported in the arthropods listed in Table II.

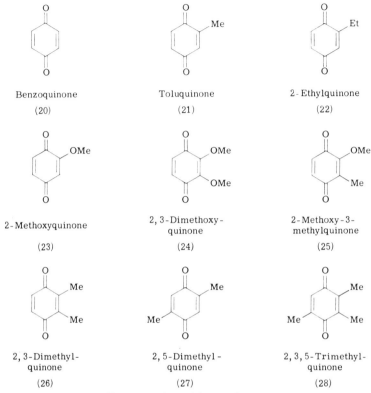

FIG. 7. Arthropod benzoquinones.

Besides being the most widely found defensive agents, quinones must rank among the oldest. In 1889 Gilson reporting on the morphology of the odoriferous glands in *Blaps mortisaga* noted that the secretion was an oil containing crystalline yellow needles. Chittenden (1896) observed that the odoriferous substance given off by even a few flour beetles (*Tribolium confusum*) was enough to impart to the flour a "persistant and disagreeable odour." Quinones in defensive secretions are probably the chief defensive agents, but it is not unusual for them to occur along with secondary compounds such as long-chain hydrocarbons or a long-chain aliphatic aldehyde or ester. The quinones are dissolved in the secondary compounds whose role is thought to be one of increasing the effectiveness of the secretion by rendering the cuticle permeable to it.

The sole arachnid which has as yet been shown to have a quinonoid defensive secretion is *Heteropachyloidellus robustus,* and from Table II it can be seen that its secretion is unique among arthropod quinones. In the early 1950's workers at the Instituto de Investigacion de Ciencies Biologicas in Montivideo, Uraguay discovered that the yellow aqueous fluid secreted by a phalangid possessed antibiotic properties. They gave the name "gonyleptidine" to this fluid and reported that it is bacteriostatic *in vitro* against Gram-positive and Gram-negative bacteria and protozoa. The identification of the active principles in gonyleptidine was then undertaken by Estable *et al.* (1955) and Fieser and Ardao (1956) and these were shown to be 2,3-dimethylbenzoquinone **(26)**, 2,5-dimethylbenzoquinone **(27)**, and 2,3,5-trimethylbenzoquinone **(28)**.

Within the class Diplopoda the orders which have been shown to secrete *p*-benzoquinones are Spirobolida, Julida, and Spirostreptida; the quinones utilized by these millipedes are benzoquinone **(20)**, toluquinone **(21)**, 2,3-dimethoxybenzoquinone **(24)**, and 2-methoxy-3-methylbenzoquinone **(25)**. Of the twenty-two species of millipedes which have been examined *Julus terrestris* and *Spirostreptus castaneus* secrete *p*-benzoquinone; three, namely, *Pachybolus laminatus, Spirostreptus virgator,* and *Rhinocricus insulatus,* secrete *p*-toluquinone as their sole quinonoid defensive substance; one, *Uroblaniulus canadensis,* secretes benzoquinone plus 2,3-dimethoxybenzoquinone, while the remaining seventeen species secrete a mixture of toluquinone and 2-methoxy-3-methylbenzoquinone. The characterization by Behal and Phisalix (1900) of the secretion from *Julus terrestris* was incomplete as a second quinone remained unidentified. In the light of the recent findings of Weatherston and Percy (1969) that the julid *Uroblaniulus canadensis* contains the new naturally occurring benzoquinone 2,3-dimethoxybenzoquinone **(24)** in addition to benzoquinone **(20)**, it is tempting to speculate that the unidentified quinone of *Julus terrestris* is 2,3-dimethoxybenzoquinone.

The use of quinonoid defensive secretions as a taxonomic tool within the orders in which they occur is not possible as a study of Table II indicates. Whereas the millipedes, in general, contain toluquinone and 2-methoxy-3-benzoquinone, the insects utilize in the main a combination of benzoquinone and toluquinone or toluquinone and 2-ethylbenzoquinone **(22)**; but within the orders, genera, and species there is no differentiation.

TABLE II

Arthropod Benzoquinones[a]

Arthropod	Compound	References
Arachnida		
Heteropachyloidellus robustus	26, 27, 28	Estable *et al.* (1955), Fieser and Ardao (1956)
Diplopoda		
Archiulus sabulosus	21, 25	Trave *et al.* (1959)
*Julus terrestris**	20	Behal and Phisalix (1900)
Trigoniulus lumbricinus	21, 25	Barbier and Lederer (1957), Monro *et al.* (1962)
Brachyiulus unlineatus	21, 25	Schildknecht and Weis (1961a)
Cylindroiulus teutonicus	21, 25	Schildknecht and Weis (1961a)
*Uroblaniulus canadensis**	20, 24	Weatherston and Percy (1969)
Chicobolus spinigerus	21, 25	Monro *et al.* (1962)
*Floridobolus penneri**	21, 25	Monro *et al.* (1962)
*Narceus annularis**	21, 25	Monro *et al.* (1962)
Narceus gordanus	21, 25	Monro *et al.* (1962)
Pachybolus laminatus	21	Barbier and Lederer (1957)
*Rhinocricus insulatus**	21	Wheeler *et al.* (1964)
Spirostreptus castaneus	20	Barbier and Lederer (1957)
*Spirostreptus virgator**	21	Barbier and Lederer (1957)
Cambala hubrichti	21, 25	Eisner *et al.* (1965)
Orthoporus flavior	21, 25	Eisner *et al.* (1965)
Orthoporus punctilliger	21, 25	Eisner *et al.* (1965)
Orthoporus conifer	21, 25	Eisner *et al.* (1965)
Doratogonus annulipes	21, 25	Eisner *et al.* (1965)
Unidentified millipedes	21, 25	Schildknecht *et al.* (1964)
Insecta		
*Diploptera punctata**	20, 21, 22	Roth and Stay (1958)
Forficula auricularia	21, 22	Schildknecht and Weis (1960)

Prionychus ater	21, 22	Schildknecht *et al.* (1964)
Brachinus crepitans	20, 21	Schildknecht (1957)
Brachinus explodens	20, 21	Schildknecht and Holoubek (1961)
Brachinus sclopeta	20, 21	Schildknecht and Holoubek (1961)
Pheropsophus catoirei	20, 21	Schildknecht and Holoubek (1961)
Clivina fossor	20, 21, 25	Schildknecht *et al.* (1968a,b)
Chlaenius vestitus	20, 21	Schildknecht *et al.* (1968a,b)
Callistus lunatus	20, 21	Schildknecht *et al.* (1968a,b)
Blaps gigas	20, 21, 22	Schildknecht *et al.* (1964)
Blaps lethifera	20, 21, 22	Schildknecht and Weis (1960), Schildknecht *et al.* (1964)
Blaps mortisaga	21, 22	Schildknecht and Weis (1960), Schildknecht *et al.* (1964)
Blaps mucronata	21, 22	Schildknecht and Weis (1960), Schildknecht *et al.* (1964)
Blaps requienii	21, 22	Schildknecht and Weis (1960), Schildknecht *et al.* (1964)
Diaperis boletti	21, 22	Schildknecht *et al.* (1964)
Diaperis maculata	21, 22	Roth and Stay (1958)
*Eleodes hispilabris**	21, 22	Blum and Crain (1961)
*Eleodes longicollis**	20, 21, 22	Chadha *et al.* (1961a)
Gnaptor spinimanus	21, 22	Schildknecht and Weis (1960)
Helops aeneus	21, 22	Schildknecht *et al.* (1964)
Helops quisquilius	21, 22	Schildknecht *et al.* (1964)
Morisia planta tingitana	21	Schildknecht and Weis (1960)
Opatroides punctulatus	21, 22	Schildknecht *et al.* (1964)
Pimelia confusa	21	Schildknecht and Weis (1960)
Scaurus uneinus	20, 21, 22	Schildknecht *et al.* (1964)
Tenebrio molitor	21	Schildknecht (1959), Schildknecht *et al.* (1964)
Tenebrio obscurus	20	Schildknecht and Weis (1960)

TABLE II (cont'd.)

Arthropod	Compound	References
Tribolium castaneum	21, 22, 23	Alexander and Barton (1943), Loconti and Roth (1963), Roth and Stay (1958)
Tribolium confusum	21, 22	Alexander and Barton (1943), Hackman et al. (1948), Engelhardt et al. (1965)
*Zophobas rugipes**	20, 21, 22	Tschinkel (1969)

[a] An asterisk indicates that the secretion contains other components.

BENZOQUINONE BIOSYNTHESIS

Most interest in the defensive benzoquinones in recent years has centered on their mode of biosynthesis. In 1948 Hackman *et al.* proposed that the defensive quinones (21,22) of *Tribolium* species arose from the same precursor as did the phenolic acids used in the tanning process by which cuticle is hardened. This scheme is outlined in Fig. 8.

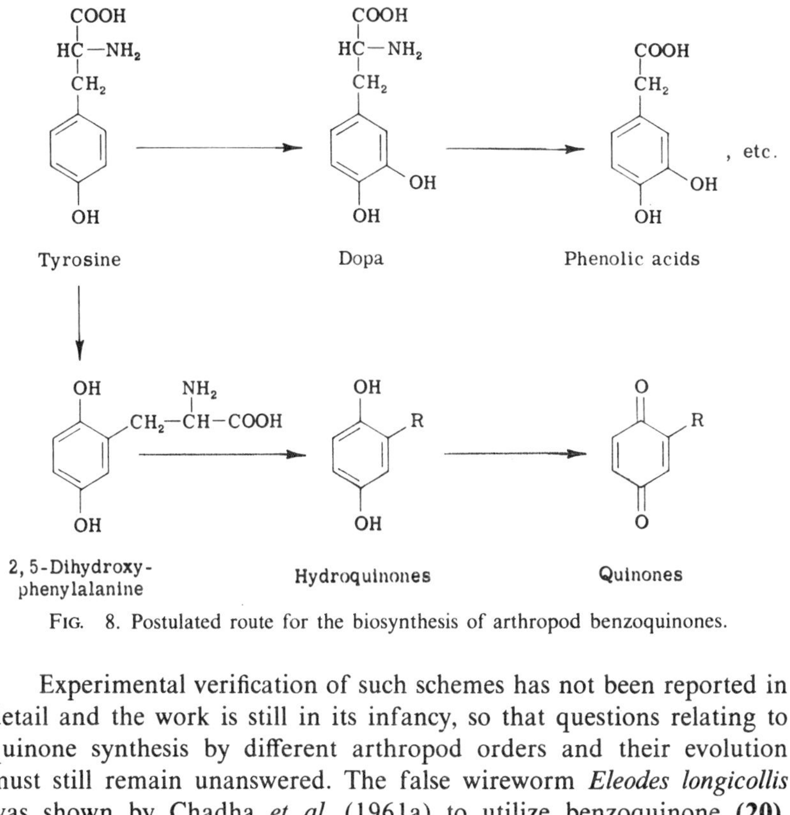

FIG. 8. Postulated route for the biosynthesis of arthropod benzoquinones.

Experimental verification of such schemes has not been reported in detail and the work is still in its infancy, so that questions relating to quinone synthesis by different arthropod orders and their evolution must still remain unanswered. The false wireworm *Eleodes longicollis* was shown by Chadha *et al.* (1961a) to utilize benzoquinone (20), toluquinone (21), and 2-ethylbenzoquinone (22). Subsequent work by Meinwald *et al.* (1966), using tracers, indicated that the quinones arise via two independent pathways. In a preliminary screening program using ^{14}C-labeled tyrosine, phenylalanine, sodium acetate, sodium propionate, and sodium malonate these authors have shown that the main pathway for the synthesis of benzoquinone involves the utilization of a preformed aromatic ring from tyrosine or phenylalanine, while incorporation in the alkylated quinones could only be rationalized by the acetate hypothesis of formation of aromatic compounds. Further experiments have defined the incorporation patterns of the alkylated quinones when 1-^{14}C-acetate and 1-^{14}C-propionate were injected into the insects. With toluquinone 30% of the activity using acetate as substrate appeared at C-2 of the ring and less than 0.1% occurred in the methyl group; hence the remaining 70% must be distributed among the other five carbon atoms of the ring. The significant result from the tagged propionate experiment was that 95% of the activity of the 2-ethylbenzoquinone was located at C-2 of the ring. These results substantiate the hypothesis that the two alkyl benzoquinones are formed by the condensation of the acyl coenzyme A and malonate units, cyclization, the formation of a quinol, and oxidation to the quinones. Quinols have been identified in several quinone-secreting insects (Schildknecht and Kramer, 1962). In the secretion of *Eleodes longicollis*, Hurst *et al.* (1964) found glucose, probably derived from the cleavage of quinol glucosides which it is reasonable to assume are intermediates in the later stages of quinone biosynthesis, in addition to the secondary components tridecane and caprylic acid. The significance of the beetle maintaining two separate pathways is not known. The large millipede *Narceus gordanus* secretes toluquinone and 2-methoxy-3-methylbenzoquinone; exploratory, tracer experiments (Weatherston and Meinwald, unpublished 1966) have shown that the methoxy group, as would be expected, is derived from methionine, 95% of the incorporated label after injection of labeled methionine being found in the methyl group of the alkoxy group. The best precursor to both quinones among the labeled substrates used was 6-methylsalicylic acid, a total of over 35% of the label being incorporated, indicating that the quinones most probably arise via the acetate pathway. A further factor complicating this study is whether the quinones are part of a sequential synthesis or whether they are formed in parallel from an as yet unknown precursor.

IV. AROMATIC ARTHROPOD DEFENSIVE COMPOUNDS

As shown in Fig. 9 there are nine aromatic compounds which serve as defensive substances; the animals from which these substances

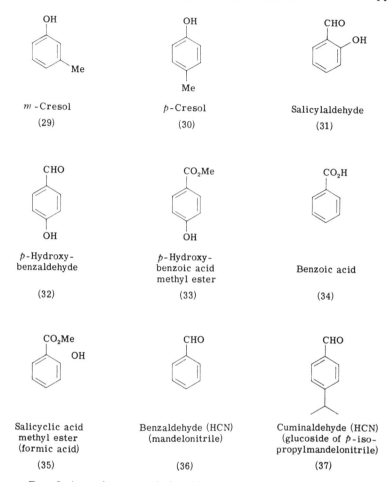

FIG. 9. Aromatic compounds found in arthropod defensive secretions.

have been isolated are given in Table III. Very little can be written in review about aromatic defensive substances and the following few paragraphs will suffice.

As stated in the footnotes to Table III members of the family Dytiscidae, the predaceous diving beetles, have evolved two types of defensive glands. One pair, situated in the prothorax, secrete steroidal compounds (these will be discussed later) and a second pair, the pygidial glands situated in the abdomen, secrete aromatic compounds. The entire work within this family has been carried out by Schild-knecht and his co-workers; for a review of this work the reader is referred to two publications from the Heidelberg group (Schildknecht *et al.*, 1964; Schildknecht, 1968). The aromatic compounds used by members of the Dytiscidae are benzoic acid (34), *p*-hydroxybenzoic acid methyl ester (33), and *p*-hydroxybenzaldehyde (32); according to German laws the first two compounds may be used in that country as preservatives in canned foods. *p*-Hydroxybenzoic acid methyl ester is widely added to synthetic diets in the culturing of insects as an antimicrobial agent. It is presumed that the insects also use these phenolic compounds as a defense against microorganisms. *p*-Hydroxy-benzaldehyde and *p*-hydroxybenzoic acid methyl ester have also been shown to be constituents of the secretion from the metathoracic defensive glands of *Ilyocoris cimicoides* (Staddon and Weatherston, 1967). This was the first report of the occurrence of aromatic compounds within the Heteroptera and once again it was presumed that the secretion was used as an antimicrobial agent. In the laboratory where *I. cimicoides* were induced to secrete, the secretion flowed on to the ventral surface of their bodies where it was distributed over the entire ventral surface by their legs; this observation endorses the assumption made above.

Two of the other aromatic aldehydes listed in Fig. 9, benzaldehyde (36) and cuminaldehyde (37), are components of one of the most interesting of all defensive secretions. The polydesmid millipedes produce cyanogenic secretions usually consisting of hydrogen cyanide and an aromatic aldehyde. Guldensteeden-Egeling (1882) was the first author to report the occurrence of hydrogen cyanide and benzaldehyde as components in a myriapode defensive secretion. Since then, this combination of defensive agents has been isolated many times. Working with *Apheloria corrugata* Eisner *et al.* (1963c) and Eisner and Eisner (1964) have shown that both components are produced from mandelo-nitrile by an elegant mechanism. In addition to the acid and aldehyde Blum and Woodring (1962) indicated the presence of glucose and a disaccharide from *Pachydesmus crassicutus,* while Barbetta *et al.* (1966) isolated D-(+)-mandelonitrile from *Gomphodesmus pavani*. According to Pallares (1946) *Polydesmus vicinus* contains the glucoside of *p*-iso-propylmandelonitrile which on enzymatic hydrolysis yields the defensive compounds hydrogen cyanide and cuminaldehyde (37) and glucose. The most interesting problem of the biosynthesis of cuminalde-hyde is as yet unresolved.

TABLE III

Aromatic Compounds Found in Arthropod Defensive Secretions[a]

Arthropod	Compound	References
Diplopoda		
Abacion magnum	30	Eisner et al. (1963a)
Apheloria corrugata*	36	H. E. Eisner et al. (1963)
Cherokia georgiana	36	H. E. Eisner et al. (1963)
Gomphodesmus pavani	36	Barbetta et al. (1966)
Pachydesmus crassicutus*	36	Blum and Woodring (1962)
Nannaria sp.	36	H. E. Eisner et al. (1963)
Polydesmus collaris collaris	36	Casnati et al. (1963)
Polydesmus gracilis*	36	Guldensteeden-Egeling (1882), Phisalix (1922)
Polydesmus vicinus*	37	Pallares (1946)
Pseudopolydesmus serratus	36	H. E. Eisner et al. (1963)
Oxidus gracilis	36	Eisner et al. (1963b)
Insecta		
Ilyocoris cimicoides	31, 32	Staddon and Weatherston (1967)
Notonecta glauca	31, 32	Pattenden and Staddon (1968)
Calosoma affini	30	McCullough (1966a)
Calosoma alternans sayi	30	McCullough (1966a)
Calosoma externum	30	McCullough and Weinheimer (1966)
Calosoma macrum	30	McCullough (1966a)
Calosoma marginalis*	30	McCullough and Weinheimer (1966)
Calosoma parvicollis	30	McCullough (1966a)
Calosoma prominens	30	Eisner et al. (1963b)
Calosoma sycophanta*	30	Casnati et al. (1965)
Bembidion quadriguttatum	31	Schildknecht et. al. (1968a)
Asaphidion flavipes	31	Schildknecht et al. (1968a)
Platynus dorsalis (Idiochroma)*	35	Schildknecht et al. (1968a,c)

TABLE III (cont'd.)

Arthropod	Compound	References
Chlaenius cordicollis	29	Eisner *et al.* (1963a)
Chlaenius chrysocephalus	29	Schildknecht *et al.* (1968a)
Chlaenius festivus	29	Schildknecht *et al.* (1968a)
Chlaenius tristus	29	Schildknecht *et al.* (1968a)
Panagaeus bipistulatus	29	Schildknecht *et al.* (1968a)
Aromia moschata	31	Hollande (1909)
Melasoma populi	31	Hollande (1909)
Phyllodecta vittelinae	31	Wain (1943)
Plagiodera sp.	31	Hollande (1909)
Cybister lateralimarginalis†	32, 33	Schildknecht *et al.* (1964), Schildknecht (1968)
Dytiscus latissimus†	32, 33, 34	Schildknecht *et al.* (1964), Schildknecht (1968)
Dytiscus marginalis†	32, 33, 34	Schildknecht *et al.* (1964), Schildknecht (1968)
Hydroporus pallustis	32	Schildknecht *et al.* (1964), Schildknecht (1968)
Zophobas rugipes†	29	Tschinkel (1969)

[a] An asterisk indicates that the secretion contains other components; a dagger indicates that the beetle contains more than one pair of defensive glands.

V. MISCELLANEOUS COMPOUNDS FOUND IN
ARTHROPOD SECRETIONS

The terpenoid, quinonoid, etc., arthropod defensive agents are well defined, but this is not the case with the miscellaneous compounds, which range from aliphatic acids and esters, through carbonyl compounds, to organic sulfides and alkaloids. The variety of compounds encompassed by this group and the large number of arthropods in which they occur make discussion extremely difficult; most of the relevant information can be obtained from Figs. 10 and 11 in conjunction with Tables IV, V, and VI.

Certain individual compounds within this broad group call for special attention. Massiolactone **(55)**, which had previously been found as a bark constituent of certain members of the laurel family (Lauraceae) and used in native medicine in New Guinea, has been isolated from several species of west Australian ants (*Camponotus* spp.). (Cavill *et al.,* 1968).

The only acetylenic compound as yet found in defensive substances has been isolated by Meinwald *et al.* (1968) from the soldier beetle *Chauliognathus lecontei*. This is the first report of dihydromatricaria acid **(56)** from arthropods, but within the plant kingdom the methyl esters of this acid and other C_{10} acetylenic acids have widespread occurrence in the Compositae.

Di- and polysulfides are known naturally as constituents of onions, garlic, etc., and recently dimethyldisulfide **(82)** and dimethyltrisulfide **(83)** were isolated from defense glands of the ponerine ant *Paltothyreus tarsatus* (Casnati *et al.,* 1967).

Glomeris marginata is a small European millipede whose defensive secretion is unique in that it contains two alkaloids. Work by Meinwald *et al.* (1966) and Schildknecht and Wenneis (1966) has shown that these alkaloids are 1,2-dimethyl-4(3)-quinazolinone [glomerin **(80)**] and 1-methyl-2-ethyl-4(3)-quinazolinone [homoglomerin **(81)**]. Using anthranilic acid with the carboxyl carbon atom labeled Schildknecht and Wenneis (1967) have further shown that this acid is a precursor to the alkaloids in the millipede.

VI. STEROIDAL DEFENSIVE COMPOUNDS

The occurrence of this well-defined class of compounds has so far been restricted to water beetles of the family Dytiscidae, and all

Aliphatic Acids and Esters

Formic acid	(38)	n-Octyl acetate	(45)
Acetic acid	(39)	trans -2-Octenyl	(46)
Caprylic acid	(40)	acetate	
Butyl butanoate	(41)	Nonyl acetate	(47)
Hexyl acetate	(42)	trans -2-Decenyl	(48)
2-Hexenyl acetate	(43)	acetate	
Hexyl butanoate	(44)	Gluconic acid	(49)

$$CH_2{=}\overset{\underset{\displaystyle CH_3}{|}}{C}-CO_2H$$

Methacrylic acid

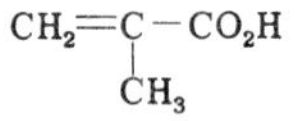

(50)

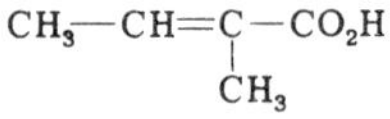

$$CH_3-CH{=}\overset{\underset{\displaystyle CH_3}{|}}{C}-CO_2H$$

Tiglic acid

(51)

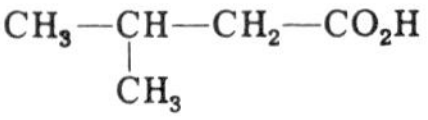

$$CH_3-\overset{\underset{\displaystyle CH_3}{|}}{CH}-CH_2-CO_2H$$

Isovaleric acid

(52)

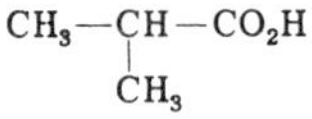

$$CH_3-\overset{\underset{\displaystyle CH_3}{|}}{CH}-CO_2H$$

Isobutyric acid

(53)

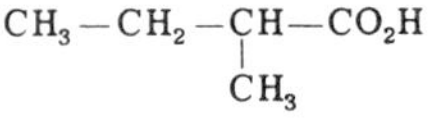

$$CH_3-CH_2-\overset{\underset{\displaystyle CH_3}{|}}{CH}-CO_2H$$

α-Methylbutyric acid

(54)

n-C_5H_{11}

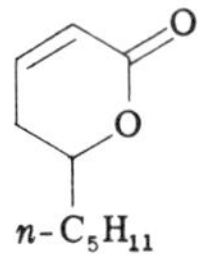

Massiolactone

(55)

$$CH_3-CH{=}CH-C{\equiv}C-C{\equiv}C-CH_2-CH_2-CO_2H$$

Dihydromatricaria acid

(56)

Carbonyl Compounds

Acetaldehyde	(57)	2-Heptanone	(68)
Butanal	(58)	4-Methyl-3-heptanone	(69)
n-Hexanal	(59)	2-Tridecanone	(70)
trans -Hexenal	(60)	2-Pentadecanone	(71)
trans -4-Oxohex-2-enal	(61)		
trans -2-Heptenal	(62)		
trans -4-Heptenal	(63)		
trans -2-Octenal	(64)		
trans -2-Decenal	(65)		
trans -2-Dodecenal	(66)		
Methyl ethyl ketone	(67)		

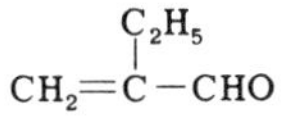

$$CH_2{=}\overset{\underset{\displaystyle CHO}{}}{\underset{|}{C}}-CHO \quad (C_2H_5)$$

2-Methylenebutenal

(72)

FIG. 10. Miscellaneous compounds found in arthropod secretions.

investigations have been carried out by Schildknecht and his associates. As noted above, members of the Dytiscidae family possess two pairs of defensive glands, the pygidial glands whose secretion contains phenolic esters and aldehydes, and the prothoracic defensive glands which secrete steroidal compounds. It is interesting to note here that *Ilyocoris cimicoides* and presumably other Naucoridae (Hemiptera), which like

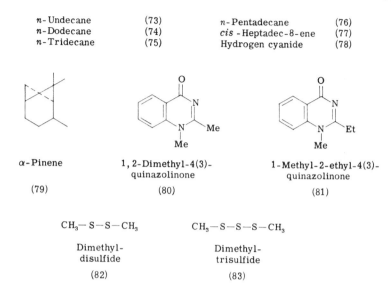

n-Undecane	(73)	*n*-Pentadecane	(76)
n-Dodecane	(74)	*cis* -Heptadec-8-ene	(77)
n-Tridecane	(75)	Hydrogen cyanide	(78)

α-Pinene

(79)

1, 2-Dimethyl-4(3)-
quinazolinone

(80)

1-Methyl-2-ethyl-4(3)-
quinazolinone

(81)

$CH_3 - S - S - CH_3$

Dimethyl-
disulfide

(82)

$CH_3 - S - S - S - CH_3$

Dimethyl-
trisulfide

(83)

FIG. 11. Other miscellaneous compounds found in arthropod secretions.

the dytiscids produce phenolic compounds, also possess prothoracic glands (Rawat, 1939). Since the diving beetles use steroidal defenses against small amphibians and fish, it is reasonable to assume that the water bugs, such as *I. cimicoides,* have also evolved a similar mechanism to repel like predators. The steroidal defense compounds and their source are given in Figs. 12 and 13 together with Table VII.

Of the eight species examined all secrete modified 4-pregnen-3-ones or 4,6-pregnadien-3-ones, including several well-known naturally occurring mammalian hormones, such as testosterone **(91)**, estrone **(95)**, and cortexone **(84)**. During their investigations on the secretion from *Acilius sulcatus* which contains components **(84–86)** Schildknecht *et al.* (1967c) examined the effect of these substances on goldfish. The mixture of steroids at a concentration of 1 mg per liter of water had almost no effect on the fish; at a concentration of 2 mg per liter of water there was considerable change in the behavior of the fish after 15 to 30 minutes. They would turn on their sides or even on their backs and could only right themselves slowly and with difficulty. After 10 to 15 minutes in water containing 10 mg per liter of steroidal mixture the goldfish were very weak and lay on the bottom of the fish tank. Occasionally this dose could prove fatal, but generally if the fish were transferred to fresh water, they would recover after ½ hour. Similar results have been obtained with testosterone (Schildknecht *et al.,* 1967a).

J. WEATHERSTON AND J. E. PERCY

TABLE IV

Aliphatic Acids and Esters Found in Arthropod Secretions[a]

Arthropod	Compound	References
Arachnida		
Mastigoproctus giganteus	39, 40	Eisner et al. (1961)
Insecta		
Agriopocoris frogatti*	42	Waterhouse and Gilby (1964)
Amorbus alternatus*	42	Waterhouse and Gilby (1964)
Amorbus rhombifer*	39, 41, 42	Waterhouse and Gilby (1964)
Amorbus rubiginosus*	42	Waterhouse et al. (1961), Waterhouse and Gilby (1964)
Aulacosternum nigrorubrum*	42	Waterhouse and Gilby (1964)
Leptocoris apicalis*	45	Baggini et al. (1966)
Mictus caja*	42	Waterhouse and Gilby (1964)
Mictus profana*	42	Waterhouse et al. (1961), Waterhouse and Gilby (1964)
Pachycolpura manca*	42	Waterhouse and Gilby (1964)
Pternistria bispina*	42, 44	Baker and Kemball (1967)
Nezara viridula var. smaragdula*	43, 46, 48	Waterhouse et al. (1961), Gilby and Waterhouse (1965)
Tessaratoma aethiops*	46	Baggini et al. (1966)
Macrocystis sp.*	46, 48	Baggini et al. (1966)
Omophron limbatum	52, 53	Schildknecht et al. (1968a)
Calosoma marginalis*	50	McCullough and Weinheimer (1966)
Calosoma scrutator	50	McCullough and Weinheimer (1966)
Calosoma sycophanta*	50, 51	Casnati et al. (1965)
Carabus auratus	50, 51	Schildknecht and Weis (1962), Schildkencht et al. (1968a)
Carabus granulatus	50, 51	Schildknecht and Weis (1962), Schildknecht et al. (1968a)
Carabus problematicus	50, 51	Schildknecht and Weis (1962), Schildknecht et al. (1968a)

Species		Reference
Cychrus rostratus	51	Schildknecht and Weis (1962)
Leistus ferrugineus	50, 51	Schildknecht *et al.* (1968a)
Nebria livida	50, 51	Schildknecht *et al.* (1968a)
Notiophilus biguttatus	52, 53	Schildknecht *et al.* (1968a)
Elaphrus ripareus	52, 53	Schildknecht *et al.* (1968a)
Lorocera pilicornis	52, 53	Schildknecht *et al.* (1968a)
Broscus cephalotes	52, 53	Schildknecht *et al.* (1968a)
Bembidion lampros	52, 53	Schildknecht *et al.* (1968a)
Bembidion andreae	52, 53	Schildknecht *et al.* (1968a)
*Calathus fuscipes**	38	Schildknecht *et al.* (1968a)
*Calathus melanocephalus**	38	Schildknecht *et al.* (1964), Schildknecht *et al.* (1968a)
*Agonum sexpunctatum**	38	Schildknecht *et al.* (1968a)
*Agonum marginatum**	38	Schildknecht *et al.* (1968a)
*Agonum viduum**	38	Schildknecht *et al.* (1968a)
*Agonum moestum**	38	Schildknecht *et al.* (1968a)
*Platynus assimilis**	38	Schildknecht *et al.* (1968a)
*Platynus dorsalis (Idiochroma)**	38	Schildknecht *et al.* (1968a,c)
*Poecilus cupreus**	50, 51	Schildknecht *et al.* (1968a)
*Pterostichus niger**	50, 51	Schildknecht and Weis (1962), Schildknecht *et al.* (1968a)
*Pterostichus macer**	50, 51	Schildknecht *et al.* (1968a)
*Pterostichus vulgaris**	50, 51	Schildknecht *et al.* (1968a)
*Pterostichus melas**	50, 51	Schildknecht *et al.* (1968a)
*Pterostichus metallicus**	50, 51	Schildknecht and Weis (1962), Schildknecht *et al.* (1968a)
Abax ater	50, 51	Schildknecht and Weis (1962), Schildknecht *et al.* (1968a)
Abax parallelus	50, 51	Schildknecht and Weis (1962), Schildknecht *et al.* (1968a)

TABLE IV (cont'd.)

Arthropod	Compound	References
Abax ovalis	50, 51	Schildknecht and Weis (1962), Schildknecht *et al.* (1968a)
Molops elatus	50, 51	Schildknecht *et al.* (1968a)
*Amara similata**	50, 51	Schildknecht *et al.* (1968a)
*Amara familiaris**	50, 51	Schildknecht *et al.* (1968a)
*Ophonus azurus**	38	Schildknecht *et al.* (1968a)
*Diachromus germanus**	38	Schildknecht *et al.* (1968a)
*Pseudophonus pubescens**	38	Schildknecht and Weis (1961b), Schildknecht *et al.* (1968a)
*Pseudophonus griseus**	38	Schildknecht and Weis (1961b), Schildknecht *et al.* (1968a)
*Anisodactylus binotatus**	38	Schildknecht and Weis (1961b), Schildknecht *et al.* (1968a)
Acinopus sp.	38	Schildknecht *et al.* (1964)
*Harpalus stratus**	38	Schildknecht *et al.* (1968a)
*Harpalus caliginosus**	38	McCullough (1966b)
*Harpalus dimidiatus**	38	Schildknecht *et al.* (1964), Schildknecht *et al.* (1968a)
*Harpalus distinguendos**	38	Schildknecht *et al.* (1968a)
*Harpalus luteicornis**	38	Schildknecht *et al.* (1968a)
*Harpalus tardus**	38	Schildknecht *et al.* (1968a)
Helluomorphoides ferrugineus	38, 47	Eisner *et al.* (1968)
Helluomorphoides latitarsus	38, 47	Eisner *et al.* (1968)
*Dichirotrichus obsoletus**	38	Schildknecht *et al.* (1968a)
*Stenolophus mixtus**	38	Schildknecht *et al.* (1968a)
*Badister bipustulatus**	38	Schildknecht *et al.* (1968a)
Dicaelus dilatatus	38	McCullough (1967a)
Dicaelus splendidus	38	McCullough (1967a)
*Licinus nitidior**	38	Schildknecht *et al.* (1968a)
Lebia chlorocephala	38	Schildknecht *et al.* (1968a)

Odacantha melanura	38	Schildknecht et al. (1968a)
Drypta dentata	38	Schildknecht et al. (1968a)
Polistichus connexus	38	Schildknecht et al. (1968a)
Chauliognathus lecontei	56	Meinwald et al. (1968)
*Eleodes longicollis**	40	Hurst et al. (1964), Meinwald and Eisner (1964)
Papilio machaon (larva)	53, 54	Eisner and Meinwald (1965)
Cerura vinula	38	Poulton (1888)
Schizura leptinoides	38	Weatherston (1967)
Many species of ants	38	Weatherston (1967) and references contained therein
*Acanthomyops claviger**	38	Regnier and Wilson (1968)
Camponotus sp.	55	Cavill et al. (1968)
*Eurycotis decipiens**	49	Dateo and Roth (1967a,b)
*Eurycotis biolleyi**	49	Dateo and Roth (1967b)
Eurycotis floridana	49	Dateo and Roth (1967b)

[a] An asterisk indicates that the secretion contains other components.

TABLE V

Carbonyl Compounds Found in Arthropod Secretions[a]

Arthropod: Insecta	Compound	References
Corixa dentipes	61	Pinder and Staddon (1965a)
Sigara falleni	61	Pinder and Staddon (1965a,b)
Cimex lectularis	57, 60, 64, 67	Schildknecht et al. (1964), Collins (1968)
Acanthocephala declivis	60	McCullough (1967b)
Acanthocephala femorata	60	Blum et al. (1961)
Acanthocephala granulosa	60	McCullough (1966c, 1967b)
Acanthocoris sordidus	59, 60	Yamamoto and Tsuyuki (1964)
Agriopocoris froggati*	59	Waterhouse and Gilby (1964)
Amorbus alternatus*	59	Waterhouse and Gilby (1964)
Amorbus rhombifer*	58, 59	Waterhouse and Gilby (1964)
Amorbus rubiginosus*	59	Waterhouse et al. (1961), Waterhouse and Gilby (1964)
Hygia opaca	59	Yamamoto and Tsuyuki (1964)
Aulacosternum nigrorubrum*	59	Waterhouse and Gilby (1964)
Leptocoris apicalis*	64, 65	Baggini et al. (1966)
Mictus caja*	59	Waterhouse and Gilby (1964)
Mictus profana*	59	Waterhouse et al. (1961), Waterhouse and Gilby (1964)
Pachycolpura manca*	59	Waterhouse and Gilby (1964)
Plinachtus bicoloripes	59	Yamamoto and Tsuyuki (1964)
Pterinistria bispina*	58, 59	Baker and Kemball (1967)
Riptortus clavatus	58	Yamamoto and Tsuyuki (1964)
Hyocephalus sp.*	59	Waterhouse and Gilby (1964)
Aelia fieberi	64, 65	Yamamoto and Tsuyuki (1964)
Biprorulus bibax*	60, 64	Park and Sutherland (1962)
Brachymena quadripustulata	60	Blum (1961)
Dolicoris baccarum*	60, 64, 65	Schildknecht et al. (1962), Schildknecht et al. (1964)
Eurygaster sp.*	60, 64	Schildknecht et al. (1964)

Euschistus servus*	62	Blum and Traynham (1960)
Graphosoma rubrolineatum	59, 65	Yamamoto and Tsuyuki (1964)
Menida scotti*	65	Yamamoto and Tsuyuki (1964)
Nezara antennata	65	Yamamoto and Tsuyuki (1964)
Nezara viridula	65	Blum and Traynham (1960), Yamamoto and Tsuyuki (1964)
Nezara viridula var. smaragdula*	60, 61, 64, 65	Waterhouse et al. (1961), Gilby and Waterhouse (1965)
Oebalus pugnax*	62	Blum and Traynham (1960), Blum et al. (1960)
Palomena viridissima*	65	Schildknecht et al. (1964)
Piezodorus teretipes*	60	Gilchrist et al. (1966)
Poecilometis strigatus*	60, 64	Waterhouse et al. (1961)
Rhoecocoris sulciventris*	60, 64	Waterhouse et al. (1961), Park and Sutherland (1962)
Scotinophara lurida*	60, 64, 65	Yamamoto and Tsuyuki (1964)
Tessaratoma aethiops*	60, 61, 64	Baggini et al. (1966)
Macrocystus sp. *	61	Baggini et al. (1966)
Scaptocoris divergens*	60, 62, 64	Roth (1961)
Unknown genera (Hemiptera: Pentatomidae)	63	Mukerji and Sharma (1966)
Dysdercus intermedius*	59, 60, 61, 64	Calam and Youdeowei (1968)
Leptoterna dolabrata*	59, 64	Collins and Drake (1965)
Cutilia soror	60	Chadha et al. (1961b)
Eurycotis biolleyi*	60	Dateo and Roth (1967b)
Eurycotis decipiens*	60	Dateo and Roth (1967b)
Eurycotis floridana*	60	Roth et al. (1956), Dateo and Roth (1967)
Pelmatosilpha coriacea	60	Blum (1964)
Platyzosteria castanea*	72	Waterhouse and Wallbank (1967)
Platyzosteria jungii*	72	Waterhouse and Wallbank (1967)
Platyzosteria morosa*	72	Waterhouse and Wallbank (1967)
Platyzosteria nova-seelandia*	60	Roth and Willis (1960)
Platyzosteria raficeps*	72	Waterhouse and Wallbank (1967)
Atta bisphaerica	68, 69	Blum et al. (1968a)

TABLE V (cont'd.)

Arthropod: Insecta	Compound	References
*Atta capiguara**	68, 69	Blum *et al.* (1968a)
Atta colombica	68, 69	Blum *et al.* (1968a)
*Atta laevigata**	68, 69	Blum *et al.* (1968a)
Atta robusta	68, 69	Blum *et al.* (1968a)
*Atta sexdens**	68, 69	Blum *et al.* (1968a)
Atta texana	68, 69	Moser *et al.* (1968)
Crematogaster africana	60	Bevan *et al.* (1961)
Pogonomyrmex badius	69	McGurk *et al.* (1966)
Pogonomyrmex barbatus	69	McGurk *et al.* (1966)
Pogonomyrmex californicus	69	McGurk *et al.* (1966)
Pogonomyrmex desertum	69	McGurk *et al.* (1966)
Pogonomyrmex occidentalis	69	McGurk *et al.* (1966)
Pogonomyrmex rugosus	69	McGurk *et al.* (1966)
*Acanthomyops claviger**	70, 71	Regnier and Wilson (1968)
*Lasius umbratus**	70	Quilico *et al.* (1957a), Blum *et al.* (1968b)
*Conomyrma pyramicus flavopectus**	68	Blum and Warter (1966), McGurk *et al.* (1968)
Iridomyrmex pruinosus	68	Blum *et al.* (1966)

[a] An asterisk indicates that the secretion contains other compounds.

TABLE VI

Other Miscellaneous Compounds Found in Arthropod Secretions[a]

Arthropod	Compound	References
Diplopoda		
Apheloria corrugata*	78	H. E. Eisner et al. (1963)
Cherokia georgiana*	78	H. E. Eisner et al. (1963)
Gomphodesmus pavani*	78	Barbetta et al. (1966)
Nannaria sp. *	78	H. E. Eisner et al. (1963)
Pachydesmus crassicutus*	78	Blum and Woodring (1962)
Polydesmus collaris collaris*	78	Casnati et al. (1963)
Polydesmus vicinus*	78	Pallares (1946)
Polydesmus virginiensis	78	H. E. Eisner et al. (1963)
Pseudopolydesmus serratus	78	H. E. Eisner et al. (1963), W. M. Wheeler (1890)
Oxidus gracilis	78	H. E. Eisner et al. (1963)
Glomeris marginata	80, 81	Meinwald et al. (1966), Schildknecht et al. (1966), Schildknecht and Wenneis (1966), Schildknecht et al. (1967)
Chilopoda		
Pachymerium ferrugineum	78	Schildknecht et al. (1968d)
Isoptera		
Nasutitermes (three species)	79	Moore (1964)
Hemiptera		
Biproulus bibax*	73, 75	Park and Sutherland (1962)
Carpocoris purpureipennis	75	Remold (1963)
Euschistus servus*	75	Blum and Traynham (1960)
Nezara viridula var. *smaragdula**	73, 74, 75	Waterhouse et al. (1961), Gilby and Waterhouse (1965)
Oebalus pugnax*	75	Blum and Traynham (1960), Blum et al. (1960)
Rhoecoris sulciventris*	75	Park and Sutherland (1962), Waterhouse et al. (1961)
Tessaratoma aethiops*	75[b]	Baggini et al. (1966)
Ceratocoris cephalicus	75	Baggini et al. (1966)

TABLE VI (cont'd.)

Arthropod	Compound	References
Macrocystus sp. *	74, 75	Baggini et al. (1966)
Dysdercus intermedius*	74, 75, 76	Calam and Youdeowei (1968)
Coleoptera		
Eleodes longicollis*	75	Meinwald and Eisner (1964), Hurst et al. (1964)
Hymenoptera		
Acanthomyops claviger*	73, 75	Regnier and Wilson (1968)
Lasius (Dendrolasius) fuliginosus*	73, 75, 76	Bernardi et al. (1967)
Paltothyreus tarsatus	82, 83	Casnati et al. (1967)
Myrmecia gulosa	77	Cavill and Williams (1967)

[a]An asterisk indicates that other compounds are found in the secretion.
[b]This compound is also found in the larva.

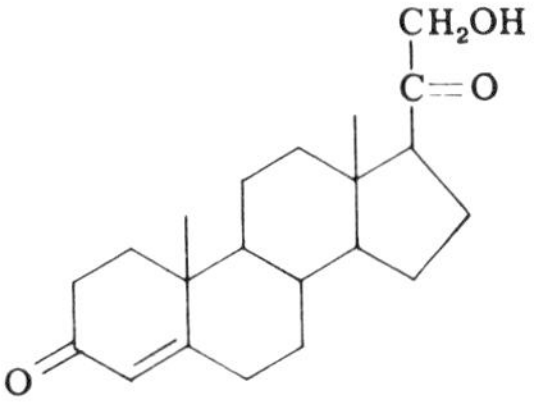

Cortexone
(84)

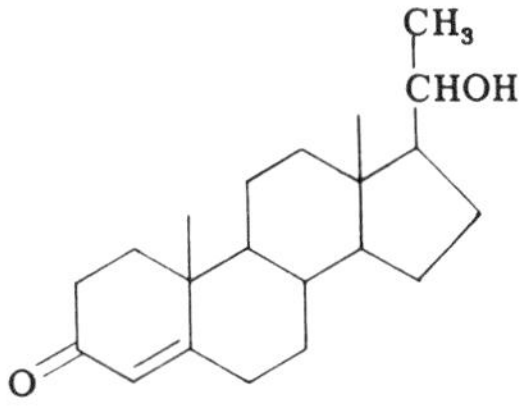

4-Pregnen-20α-ol-3-one
(85)

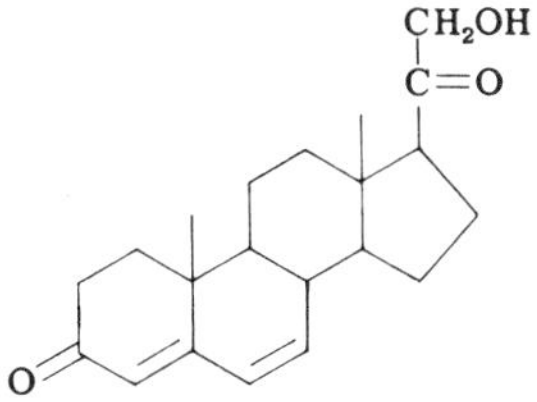

4,6-Prenadien-
21-ol-3, 20-dione
(86)

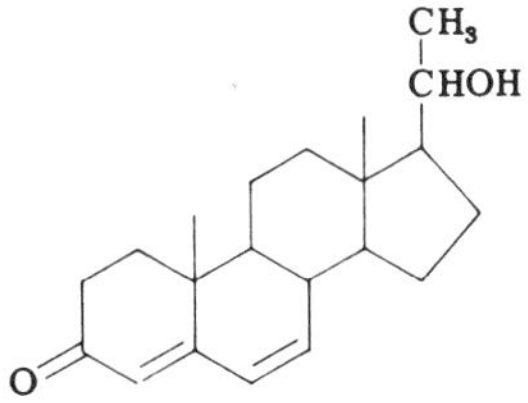

4,6-Pregnadien-20α-
ol-3-one (cybisterone)
(87)

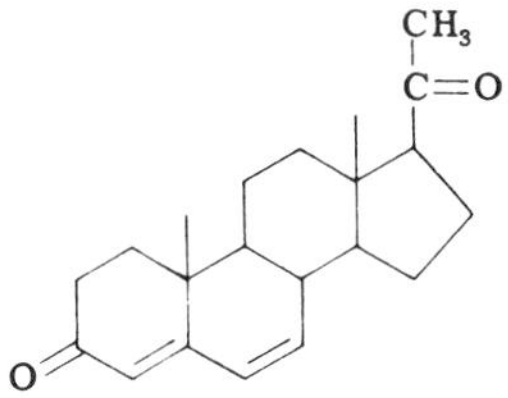

4,6-Pregnadiene-
3, 20-dione
(88)

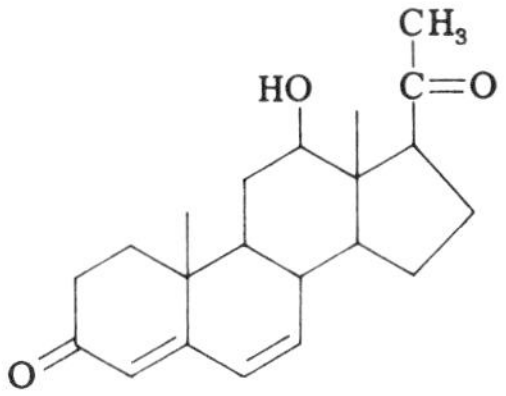

12β-Hydroxy-4, 6-pregnadiene-
3, 20-dione (cybisterol)
(89)

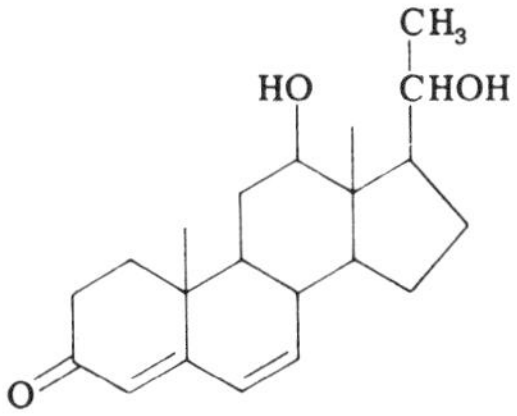

12β-Hydroxycybisterone
(90)

FIG. 12. Defensive steroidal compounds isolated from beetles of the family Dytiscidae.

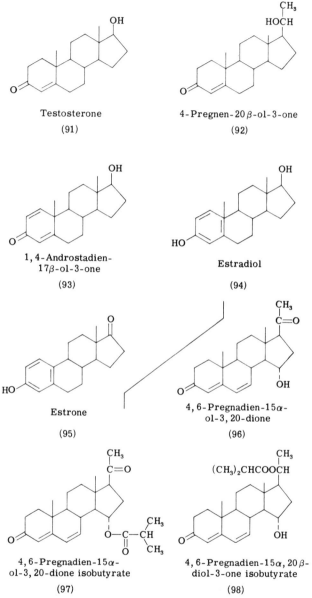

Testosterone
(91)

4-Pregnen-20β-ol-3-one
(92)

1,4-Androstadien-
17β-ol-3-one
(93)

Estradiol
(94)

Estrone
(95)

4,6-Pregnadien-15α-
ol-3,20-dione
(96)

4,6-Pregnadien-15α-
ol-3,20-dione isobutyrate
(97)

4,6-Pregnadien-15α,20β-
diol-3-one isobutyrate
(98)

FIG. 13. Defensive steroidal compounds isolated from *Ilybius fenestratus* and *Agabus sturmi* (Dytiscidae).

TABLE VII

Steroids in Dytiscid Prothoracid Defensive Glands

Insect, Coleoptera: Dytiscidae	Compound	References
Agabus bipustulatus	84	Schildknecht (1968)
Agabus sturmi	96, 97, 98	Schildknecht (1968)
Ilybius fenestratus	91, 92, 93, 94, 95	Schildknecht *et al.* (1967a), Schildknecht (1968)
Ilybius fuliginosus	91	Schildknecht *et al.* (1967a)
Cybister lateralimarginalis	86, 87, 88, 90	Schildknecht *et al.* (1967b), Schildknecht (1968)
Cybister tripunctatus	87, 89	Schildknecht (1968), Schildknecht and Kornig (1968)
Dyticus marginalis	84, 85, 87	Schildknecht and Maschwitz (1966), Schildknecht and Hotz (1967), Schildknecht (1968)
Acilius sulcatus	84, 85, 86, 87, 88	Schildknecht *et al.* (1967c), Schildknecht (1968)

VII. NONEXOCRINE DEFENSIVE COMPOUNDS

As stated in the introduction, our knowledge of nonexocrine defensive substances is far less extensive than of their exocrine counterparts, but these substances are no less important. The chemical structure of these nonexocrine agents are usually more complex than the defensive substances already discussed, and from Fig. 14 it can be seen that several of them are of pharmacological significance. The whole question concerning this type of defensive agent is more complex. As has already been pointed out in the cases of the quinones and the ant terpenes, the active compounds are biosynthesized by the secreting organism, and the same compound may be secreted by members of different genera, families, and even orders. On the other hand the nonexocrine materials are generally restricted to members of the same genera and are usually obtained, albeit in a slightly modified form, from the food plant of the insect. For example, Rothschild *et al.* (1968) have reported the presence of the nitrophenanthrene derivative aristolochic acid **(103)** in the body tissues of the swallow-tail butterfly *Papilio aristolochiae*. Aristolochic acid occurs in many species of plants belonging to the family Aristolochiaceae.

Table VIII indicates the arthropods which have been shown to possess nonexocrine defensive substances.

Cantharidin, the defensive agent of many species of blister beetles has been widely investigated, in part no doubt due to the notoriety it has received as an aphrodisiac called "Spanish Fly." Because of its toxicity, it can cause gastroenteritis, renal damage, haematuria, and even death. It is now of little medicinal value, although it was previously used widely as a vesicant. The structure **(99)** was proposed by Gadamer (1914) and Rudolph (1916), substantiated by the synthesis of deoxycantharidin by Woodward and Loftfield (1941), and verified by total synthesis in 1953 (Stork *et al.,* 1953).

Until recently the only known cardiac glycosides known to occur in the animal kingdom were the bufadienolides or toad poisons; however, recently it has been shown that two species of danaiid butterflies (*Danaus plexippus* and *D. chrysippus*) and several species of grasshoppers of the genera *Poekilocerus* and *Phymateus* store cardiac glycosides as a means of defense. Although the animals are unrelated, both the grasshoppers and the larval form of the butterflies feed exclusively on plants of the milkweed family (Asclepiadaceae) that are known to

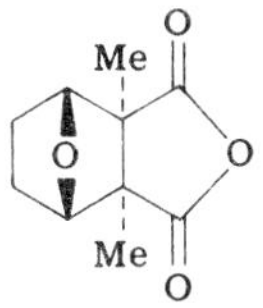

Cantharidin
(99)

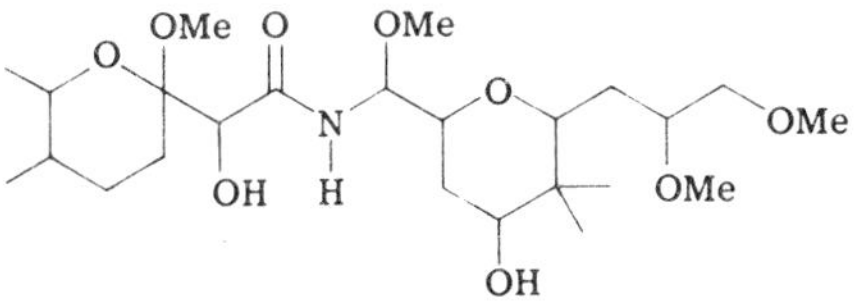

Pederin
(100)

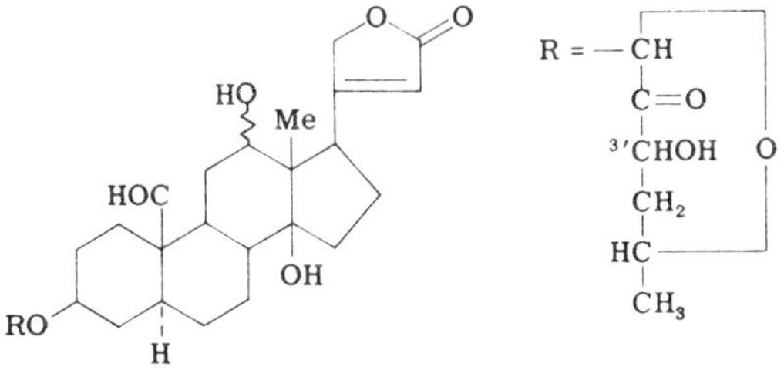

Calactin and calotropin
(101)

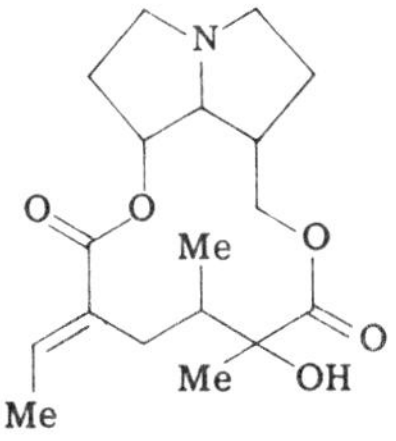

Senecionine
(102)

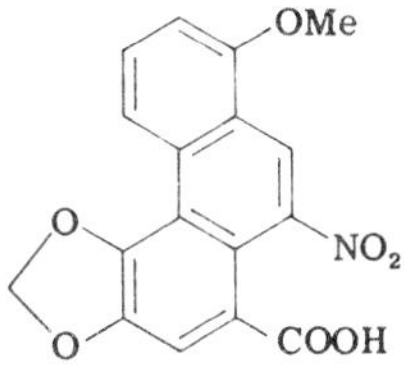

Aristolochic acid
(103)

Hydrogen cyanide
(78)

FIG. 14. Nonexocrine defensive secretions.

contain cardiac glycosides. The food plants may contain up to eight or nine cardiac glycosides, but the main glycosides stored by the butterflies and the grasshoppers are calactin and calotropin. These are both represented in Fig. 14 by formula (101) since they are thought to be isomeric at the 3'-position in the sugar moiety (Reichstein, 1967). By rearing *Poekilocerus bufonius* on nonpoisonous plants Von Euw *et al.* (1967) reduced the cardiac glycoside content by a factor of ten, and

TABLE VIII

Nonexocrine Defensive Secretions

Insect	Compound	References
Orthoptera		
Phymateus bacatuss	101	von Euw *et al.* (1967), Reichstein (1967)
Phymateus viridipes	101	von Euw *et al.* (1967), Reichstein (1967)
Poekilocerus bufonius	101	von Euw *et al.* (1967), Reichstein (1967)
Poekilocerus pictus	101	von Euw *et al.* (1967), Reichstein (1967)
Lepidoptera		
Callimorpha jacobaeae	102	Aplin *et al.* (1968)
Procoris geryon	78	Jones *et al.* (1962)
Zygaena filipendulae	78	Jones *et al.* (1962)
Zygaena lonicerae	78	Jones *et al.* (1962)
Zygaena trifolii	78	Jones *et al.* (1962)
Amauris echeria lobengula	101	Reichstein (1967), Reichstein *et al.* (1968)
Danaus chrysippus	101	Reichstein (1967), Reichstein *et al.* (1968)
Danaus plexippus	101	Reichstein (1967), Reichstein *et al.* (1968)
Papilio aristolochiae	103	Rothschild *et al.* (1968)
Coleoptera		
Fam. Meloidae (several species)	99	Walter and Cole (1967) and references contained therein
Paederus fuscipes	100	Cardani *et al.* (1966), Matsumoto *et al.* (1968) and references contained therein

second generation animals contained a quantity further reduced by a factor of seven. The residual heart poisons are attributed to the fact that those fed on a nonpoisonous diet were not reared in isolation and grasshoppers are known to consume each others' exuviae, which were shown to be rich in cardenolides.

Aposematic moths of the genus *Zygaena* have been shown to contain hydrogen cyanide (Jones *et al.*, 1962), but whether this toxin is obtained from a food plant source has not been established. On the other hand, the larvae of the cinnabar moth (*Callimorpha jacobaeae*) feed on groundsel (*Senecio vulgaris*) and ragwort (*Senecio jacobaeae*), and it is thought that the *Senecio* alkaloids retained in the body tissues of the moths play some part in their defensive mechanism (Aplin *et al.*, 1968). Several alkaloids are stored by the moths, senecionine **(102)** being present in the greatest quantity.

The most complex of all the insect defensive materials is that from the staphylinid beetle *Paederus fuscipes*. In 1952 Pavan and Bo reported the isolation of a toxic principle from *P. fuscipes* to which they gave the name pederin. An elegant review of the biological and medicinal researches was published by Pavan (1963). Chemical investigations of the Italian group led to the elucidation of the structure in 1966 (Cardani *et al.*, 1966); their structure, however, has been recently modified to **(100)** on spectroscopic evidence and X-ray crystallography (Matsumoto *et al.*, 1968; Furusaki *et al.*, 1968). A third active material, named pederone, was recently reported (Cardani *et al.*, 1967); in view of the recent Japanese work the structure of this compound is presumably **(100)** modified by having a carbonyl group replace the hydroxyl at C-4.

VIII. QUESTIONS UNANSWERED

As noted by Eisner and Meinwald (1966) the area of arthropod defensive substances is far from exhausted and should blossom for many years to come. Hence it seems appropriate to conclude this brief review by mentioning a few of the unanswered questions.

The study of the biosynthesis of the secretions is in its infancy and the formation of aromatic compounds from acetate units, a process until recently thought not to occur in insects, deserves further investigation. Biosynthetic studies in regard to the vertebrate hormones and other steroidal compounds isolated from the dytiscid beetles should prove fruitful in view of the currently held hypothesis of the limited ability of arthropods to synthesize such compounds from nonsteroid precursors.

The knowledge of the morphology, histology, and histochemistry of the glands producing the active agents needs to be greatly increased in order that an insight might be gained into the question of how the gland cells manufacture such poisons without any injury to the parent animal. In a similar vein Eisner and Meinwald (1966) report that the millipede *Apheloria corrugata,* which utilizes hydrogen cyanide as one of its defensive secretions, usually outlives other arthropods confined with it in a cyanide killing jar. As yet very little is known about the cyanide detoxification mechanism utilized by these millipedes.

Other general areas which should be explored are the effect of the secretions against microbes and entomophageous parasites and finally their medicinal and insecticidal applications.

The above paragraphs do not exhaust the long list of unanswered questions, but give an insight to the vast amount of work still to be done in this most interesting area of insect chemistry.

SUMMARY

The literature concerning defensive mechanisms of arthropods has been reviewed; the compounds used for defense and the species from which they originate are presented in tabular form. The substances have been arbitrarily classified into *(a)* terpenoid and related compounds, *(b)* benzoquinones, *(c)* aromatic compounds, *(d)* miscellaneous compounds including aliphatic hydrocarbons, acids, esters, and carbonyl compounds, *(e)* steroids, and *(f)* nonexocrine secretions. Aspects of the chemistry and biosynthesis of these compounds are presented. Several current unsolved problems are briefly discussed. The literature survey pertaining to this review was concluded in January, 1969.

REFERENCES

Alexander, P., and Barton, D. H. R. (1943). The excretion of ethylquinone by the flour beetle. *Biochem. J.* **37**, 463.

Aplin, R. T., Benn, M. H., and Rothschild, M. (1968). Poisonous alkaloids in the body tissue of the cinnabar moth *(Callimorpha jacobaeae). Nature* **219**, 747.

Baggini, A. R., Bernardi, R., Casnati, G., Pavan, M., and Ricca, A. (1966). Ricerche sulle secrezioni defensive di Insetti Emitteri Eterotteri. *Rev. Espan. Entomol.* 42, 7.

Baker, J. T., and Kemball, P. A. (1967). Volatile constituents of the scent gland reservoir of the coreoid *Pternistria bispina* (Hemiptera). *Australian J. Chem.* 20, 395.

Barbetta, M., Casnati, G., and Pavan, M. (1966). The presence of D(+) mandelonitrile in the defensive secretion of the myriapode *Gomphodesmus pavani.* Memoire della societa entomol. *Italiana* 45, 169.

Barbier, M., and Lederer, E. (1957). Sur les benzoquinones du venim de troi especes de Myriapodes. *Biokimiya* 22, 236.

Behal, A., and Phisalix, M. (1900). La quinone, principe actif du venim du *Julus terrestris. Bull. Museum Natl. Hist. Nat. (Paris)* 6, 338.

Bernardi, R., Cardani, C., Ghiringhelli, D., Selva, A., Baggini, A., and Pavan, M. (1967). On the components of secretion of mandibular glands of the ant *Lasius (Dendrolasius) fuliginosus. Tetrahedron Letters* p.3893.

Bevan, C. W. L., Birch, A. J., and Caswell, H. (1961). Insect repellent from black cocktail ants. *J. Chem. Soc.* p.488.

Blum, M. S. (1961). The presence of 2-hexenal in the scent gland of the pentatomid *Brachymena quadripustulata. Ann. Entomol. Soc. Am.* 54, 410.

Blum, M. S. (1964). Insect defensive secretion: Hex-2-enal-l in *Pelmatosilpha coriacea* (Blattaria) and its repellent value under natural conditions. *Ann. Entomol. Soc. Am.* 57, 600.

Blum, M. S. (1969). Alarm pheromones. *Ann. Rev. Entomol.* 14, 57.

Blum, M. S., and Crain, R. D. (1961). The occurrence of para-quinones in the abdominal secretions of *Eleodes hispilabris* (Coleoptera: Tenebrionidae). *Ann. Entomol. Soc. Am.* 54, 474.

Blum, M. S., and Traynham, J. G. (1960). The chemistry of the pentatomid scent gland. *Proc. XIth Intern. Congr. Entomol., Vienna, 1960* Vol. III, p.48.

Blum, M. S., and Warter, S. L. (1966). Chemical releasers of social behaviour. VII. The isolation of 2-heptanone from *Conomyrma pyramicus* (Hymenoptera: Formicidae) (Dolichoderinae) and its modus operandi as a releaser of alarm and digging behaviour. *Ann. Entomol. Soc. Am.* 59, 774.

Blum, M. S., and Woodring, J. P. (1962). Secretion of benzaldehyde and hydrogen cyanide by the millipede *Pachydesmus crassicutus* (Wood). *Science* 138, 512.

Blum, M. S., Traynham, J. G. Chidester, J. B., and Boggus, J. D. (1960). *n*-Tridecane and *trans*-2-heptenal in scent gland of the rice stink bug, *Oebalus pugnax* (F.). *Science* 132, 1480.

Blum, M. S., Crain, R. D., and Chidester, J. B. (1961). Trans-2-hexenal in the scent gland of the Hemipteran *Acanthocephala femorata. Nature* 189, 245.

Blum, M. S., Warter, S. L., and Traynham, J. G. (1966) Chemical releasers of social behaviour. VI. The relation of structure to the activity of ketones as releasers of alarm for *Iridomyrmex pruinosus. J. Insect Physiol.* 12, 419.

Blum, M. S., Padovani, F., and Amante, E. (1968a). Alkanones and terpenes in the mandibular glands of *Atta* species. (Hymenoptera: Formicidae). *Comp. Biochem. Physiol.* 26, 291.

Blum, M. S., Padovani, F., Hermann, H. R., Jr., and Kannowski, P. B. (1968b). Chemical releasers of social behaviour. XI. Terpenes in the mandibular glands of *Lasius umbratus. Ann. Entomol. Soc. Am.* 61, 1354.

Butenandt, A. (1959). Wirkstoffe des Insektenreiches. *Naturwissenschaften* 46, 461.

Butenandt, A., Linzen, B., and Lindauer, M. (1959). Uber einen Duftstoff aus der

Mandibeldrüse der Blattschnie der Ameise *Atta sexdens rubropilosa* (Forel). *Arch. Anat. Microscop. Morphol. Exptl.* **48**, 13.

Butler, C. G. (1967). Insect pheromones. *Biol. Rev.* **42**, 42.

Calam, D. H., and Youdeowei, A. (1968). Identification and functions of secretions from the posterior scent gland of the fifth instar larva of the bug *Dysdercus intermedius. J. Insect Physiol.* **14**, 1147.

Cardani, C., Ghiringhelli, D., Mondelli, R., and Quilico, A. (1966). Struttura della pederina. *Gazz. Chim. Ital.* **96**, 3.

Cardani, C., Ghiringhelli, D., Quilico, A., and Selva, A. (1967). The structure of pederone a novel substance from *Paederus fuscipes* (Coleoptera Staphylinidae). *Tetrahedron Letters* p.4023.

Casnati, G., Nencini, G., Quilico, A., Pavan, M., Ricca, A., and Salvatori, T. (1963). The secretion of the myriapod *Polydesmus collaris collaris* (Koch). *Experientia* **19**, 409.

Casnati, G., Pavan, M., and Ricca, A. (1965). Sulla costituzione del veleno dell'insetto *Calosoma sycophanta* L. (Coleoptera Carabidae). *Ann. Soc. Entomol. France* [N.S.] **1**, 705.

Casnati, G., Ricca, A., and Pavan, M. (1967). Defensive secretion of the mandibular glands of *Paltothyreus tarsatus. Chim. Ind. (Milan)* **49**, 57.

Cavill, G. W. K. (1960). The cyclopentanoid monoterpenes. *Rev. Pure Appl. Chem.* **10**, 169.

Cavill, G. W. K., and Clark, D. V. (1967). Insect venoms, attractants, and repellents. VIII. Isodihydronepetalactone. *J. Insect Physiol.* **13**, 131.

Cavill, G. W. K., and Hinterberger, H. (1960). Chemistry of ants; IV. Terpenoid constituents of some *Dolichoderus* and *Iridomyrmex* species. *Australian J. Chem.* **13**, 514.

Cavill, G. W. K., and Robertson, P. L. (1965). Ant venoms, attractants, and repellents. *Science* **149**, 1337.

Cavill, G. W. K., and Williams, P. J. (1967). Constituents of Dufours glands in *Myrmecia gulosa. J. Insect Physiol.* **13**, 1097.

Cavill, G. W. K., Ford, D. L., and Locksley, H. O. (1956). The chemistry of ants. I. Terpenoid constituents of some Australian *Iridomyrmex* species. *Australian J. Chem.* **9**, 288.

Cavill, G. W. K., Williams, P. J., and Whitfield, F. B. (1967). α-Farnesene, Dufours gland secretion in *Aphaenogaster longiceps* (F.Sm.). *Tetrahedron Letters* p.2201.

Cavill, G. W. K., Clark, D. V., and Whitfield, F. B. (1968). Insect venoms, attractants, and repellents. XI. Massiolactone from two species of formicine ants, and some observations on constituents of the bark oil of *Cryptocarya massoia. Australian J. Chem.* **21**, 2819.

Chadha, M. S., Eisner, T., and Meinwald, J. (1961a). Defense mechanisms of arthropods. IV. Para-benzoquinones in the secretion of *Eleodes longicollis* Lec. (Coleoptera: Tenebrionidae). *J. Insect Physiol.* **7**, 46.

Chadha, M. S., Eisner, T., and Meinwald, J. (1961b). Defense mechanisms of arthropods. III. Secretion of 2-hexenal by adults of the cockroach *Cutilia soror* (Brunner). *Ann. Entomol. Soc. Am.* **54**, 642.

Chadha, M. S., Eisner, T., Monro, A., and Meinwald, J. (1962). Defense mechanisms of arthropods. VII. Citronellal and citral in the mandibular gland secretion of the ant *Acanthomyops claviger. J. Insect Physiol.* **8**, 175.

Chittenden, F. H. (1896). Insects affecting cereals and other dry vegetable foods. *U.S. Dept. Agr. Entomol. Bull.* **4**, 112.

Collins, R. P. (1968). Carbonyl compounds produced by the bedbug, *Cimex lectularis. Ann. Entomol. Soc. Am.* **61**, 1338.

Collins, R. P., and Drake, T. H. (1965). Carbonyl compounds produced by the meadow plant bug *Leptoterna dolabrata* (Hemiptera: Meridae). *Ann. Entomol. Soc. Am.* **58**, 764.

Dateo, G. P., and Roth, L. M. (1967a). D-gluconic acid: Isolation from the defensive secretion of the cockroach *Eurycotis decipiens. Science* **155**, 88.

Dateo, G. P., and Roth, L. M. (1967b). Occurrence of gluconic acid and 2-hexenal in the defensive secretions of three species of *Eurycotis* (Blattaria:Blattidae:Polyzosteriinae). *Ann. Entomol. Soc. Am.* **60**, 1025.

Eisner, H. E., Eisner, T., and Hurst, J. J. (1963). Hydrogen cyanide and benzaldehyde produced by millipedes. *Chem. Ind. (London)* p.124.

Eisner, T., and Eisner, H. E. (1964). Mystery of a millipede. *Nat. Hist.* **74**, 30.

Eisner, T., and Meinwald, Y. C. (1965). Defensive secretion of a caterpillar *(Papilio). Science* **150**, 1733.

Eisner, T., and Meinwald, J. (1966). Defensive secretions of arthropods. *Science* **153**, 1341.

Eisner, T., Meinwald, J., Monro, A., and Ghent, R. (1961). Defense mechanisms of arthropods. I. The composition and function of the spray of the whipscorpion, *Mastigoproctus giganteus* (Lucas) (Arachnida: Pedipalpida). *J. Insect Physiol.* **6**, 272.

Eisner, T., Hurst, J. J., and Meinwald, J. (1963a). Defense mechanisms of arthropods. XI. The structure, function, and phenolic secretions of the glands of a chordeumoid millipede and carabid beetle. *Psyche* **70**, 94.

Eisner, T., Swithenbank, C., and Meinwald, J. (1963b). Defense mechanisms of arthropods. VIII. Secretion of salicylaldehyde by a carabid beetle. *Ann. Entomol. Soc. Am.* **56**, 37.

Eisner, T., Eisner, H. E., Hurst, J. J., Kafatos, F. C., and Meinwald, J. (1963c). Cyanogenic glandular apparatus of a millipede. *Science* **139**, 1218.

Eisner, T., Hurst, J. J., Keeton, W. T., and Meinwald, Y. C. (1965). Defense mechanisms of arthropods. XVI. Para-benzoquinones in the secretion of spirostreptoid millipedes. *Ann. Entomol. Soc. Am.* **58**, 247.

Eisner, T., Meinwald, Y. C., Alsop, D. W., and Carrel, J. E. (1968). Defense mechanisms of arthropods. XXI. Formic acid and *n*-nonyl acetate in the defensive spray of two species of *Helluomorphoides. Ann. Entomol. Soc. Am.* **61**, 610.

Engelhardt, M., Rapaport, H., and Sokoloff, A. (1965). Odorous secretion of normal and mutant *Tribolium confusum. Science* **150**, 632.

Estable, C., Ardao, M. I. Brasil, N. P., and Fieser, L. F. (1955). Gonyleptidine. *J. Am. Chem. Soc.* **77**, 4942.

von Euw, J., Fishelson, L., Parsons, J. A., Reichstein, T., and Rothschild, M. (1967). Cardenolides (heart poisons) in a grasshopper feeding on milkweeds. *Nature* **214**, 35.

Fieser, L. F., and Ardao, M. I. (1956). Investigation of the chemical nature of Gonyleptidine. *J. Am. Chem. Soc.* **78**, 744.

Fisher, S. (1670). Reported by J. Wray. Some uncommon observations and experiments made with an acid juice to be found in ants. *Phil. Trans. Roy. Soc. (London)* p.2063.

Furusaki, A., Watanabe, T., Matsumoto, T., and Yanagiya, M. (1968). The crystal and molecular structure of pederin di-*p*-bromobenzoate. *Tetrahedron Letters,* p.6301.

Gadamer, J. (1914). The constitution of cantharidin. *Arch. Pharm.* **252**, 609.

Gilby, A. R., and Waterhouse, D. F. (1965). The composition of the scent of the green vegetable bug, *Nezara viridula. Proc. Roy. Soc. (London).* **B162**, 105.

Gilchrist, T. L., Stansfield, F., and Cloudsley-Thompson, J. L. (1966). The odoriferous principle of *Piezodorus teretipes* (Stol) (Hemiptera: Pentatomoidae). *Proc. Roy. Entomol. Soc. (London)* **A41**, 55.

Gilson, G. (1889). Les glandes odoriferes du *Blaps mortisaga* et de quelques autre espèces. *Cellule* **5**, 3.

Grunanger, P., Quilico, A., and Pavan, M. (1960). Sul secreto odoroso del formicide *Myrmecaria natalensis* (Fred). *Rend. Accad. Sci. Fis. Nat. (Soc. Nazl. Sci. Napoli)* **28**, 293.

Guldensteeden-Egeling, C. (1882). Uber Bildung von Cyanwasserstoffsaure bei einem myriapoden. *Arch. Ges. Physiol.* **28**, 576.

Hackman, R. H., Pryor, M. G. M., and Todd, A. R. (1948). The occurrence of phenolic substances in arthropods. *Biochem. J.* **43**, 474.

Happ, G. M., and Meinwald, J. (1965). Biosynthesis of arthropod secretions Pt. 1. Monoterpene biosynthesis in an ant *Acanthomyops claviger*. *J. Am. Chem. Soc.* **87**, 2507.

Happ, G. M., and Meinwald, J. (1966). Biosynthesis of monoterpenes in an ant *(Acanthomyops claviger)*. *Advan. Chem. Ser.* **53**, 27.

Hollande, A-Ch. (1909). Sur la fonction d'excrétion chez les insectes salicioles et en particulier sur l'existence des dérivés salicylés. *Ann. Univ. Grenoble, Sect. Sci-Med.* **21**, 459.

Hurst, J. J., Meinwald, J., and Eisner, T. (1964). Defense mechanisms of arthropods. XII. Glucose and hydrocarbons in the quinone-containing secretion of *Eleodes longicollis*. *Ann. Entomol. Soc. Am.* **57**, 44.

Jacobson, M. (1965). "Insect Sex Attractants." (Interscience), Wiley New York.

Jacobson, M. (1966). Chemical insect attractants and repellents. *Ann. Rev. Entomol.* **11**, 403.

Jones, D. A., Parsons, J., and Rothschild, M. (1962). Release of hydrogen cyanide from the crushed tissues of all stages in the life cycle of the Zygaeninae. *Nature* **193**, 52.

Karlson, P., and Butenandt, A. (1959). Pheromones (Ectohormones) in insects. *Ann. Rev. Entomol.* **4**, 39.

Kistner, D. H., and Blum, M. S. (1969). Unpublished observations From Blum, M. S. *Ann. Rev. Entomol.* **14**, 57.

Loconti, J. D., and Roth, L. M. (1953). Composition of the odorous secretion of *Tribolium castaneum. Ann. Entomol. Soc. Am.* **46**, 281.

McCullough, T. (1966a). Quantitative determination of salicylaldehyde in the scent fluid of *Calosoma macrum, C. alternans sayi, C. affini,* and *C. parvicollis* (Coleoptera: Carabidae). *Ann. Entomol. Soc. Am.* **59**, 1018.

McCullough, T. (1966b). Compounds in the defensive scent glands of *Harpalus caliginosus* (Coleoptera: Carabidae). *Ann. Entomol. Soc. Am.* **59**, 1020.

McCullough, T. (1966c). Carbonyl and acid compounds produced by *Acanthocephala granulosa* (Hemiptera:Coreidea). *Ann. Entomol. Soc. Am.* **59**, 410.

McCullough, T. (1967a). Compounds found in the defensive scent fluid of *Dicaelus splendidus* and *D. dilatatus. Ann. Entomol. Soc. Am.* **60**, 861.

McCullough, T. (1967b). Quantitative determination of trans-2-hexenal in *Acanthocephala declivis* and *A. granulosa. Ann. Entomol. Soc. Am.* **60**, 862.

McCullough, T., and Weinheimer, A. J. (1966). Compounds found in the defensive scent fluids of *Calosoma marginalis, C. externum,* and *C. scrutator. Ann. Entomol. Soc. Am.* **59,** 410.

McGurk, D. J., Frost, J., Eisenbraun, E. J., Vick, K., Drew, W. A., and Young, J. (1966). Volatile compounds in ants: identification of 4-methyl-3-heptanone from *Pogomyrmex* ants. *J. Insect Physiol.* **12,** 1435.

McGurk, D. J., Frost, J., Waller, G. R., Eisenbraun, E. J., Vick, K., Drew, W. A., and Young, J. (1968). Iridodial isomer variation in Dolichoderine ants. *J. Insect Physiol.* **14,** 841.

Matsumoto, T., Yanagiya, M., Maeno, S., and Yasuda, S. (1968). A revised structure of pederin. *Tetrahedron Letters* p.6297.

Meinwald, J., Meinwald, Y. C., Chalmers, A. M., and Eisner, T. (1968). Dihydromatricaria acid: Acetylenic acid secreted by soldier beetle. *Science* **160,** 890.

Meinwald, Y. C., and Eisner, T. (1964). Defense mechanisms of arthropods. XIV. Caprylic acid: an accessory component of the secretion of *Eleodes longicollis. Ann. Entomol. Soc. Am.* **57,** 513.

Meinwald, Y. C., Meinwald, J., and Eisner, T. (1966). 1,2-Dialkyl-4(3H)-quinazolinones in the defensive secretion of a millipede *(Glomeris marginata). Science* **154,** 390.

Melander, A. I., and Brues, O. T. (1906). The chemical nature of some insect secretions. *Bull. Wisconsin Nat. Hist. Soc.* **4,** 22.

Monro, A., Chadha, M., Meinwald, J., and Eisner, T. (1962). Defense mechanisms of arthropods. VI. Paraquinones in the secretion of five species of millipedes. *Ann. Entomol. Soc. Am.* **55,** 261.

Moore, B. P. (1964). Volatile terpenes from *Nasutitermes* soldiers (Isoptera:Termitidae). *J. Insect Physiol.* **10,** 371.

Moser, J. C., Brownlee, R. C., and Silverstein, R. (1968). Alarm pheromones of the ant *Atta texana. J. Insect Physiol.* **14,** 529.

Mukerji, S. K., and Sharma, H. L. (1966). Investigations on the offensive odour of Hemiptera. *Tetrahedron Letters* p.2479.

Pallares, E. S. (1946). Notes on the poison produced by the *Polydesmus (Fontaria) vicinus* (L.). *Arch. Biochem.* **9,** 105.

Park, R. J., and Sutherland, M. D. (1962). Volatile constituents of the bronze orange bug, *Rhoecocoris sulciventris. Australian J. Chem.* **15,** 172.

Pattenden, G., and Staddon, B. W. (1968). Secretion of the metathoracic glands of the water bug *Notonecta glauca* (Heteroptera: Notonectidae). *Experientia* **24,** 1092.

Pavan, M. (1956). M. studi sui Formicidae. II. Sull origine, significato biologico e isolamento della dendrolasina. *Ric. Sci.* **26,** 144.

Pavan, M. (1963). Ricerche biologiche e mediche su pederina e su estratti purificati di *Paederus fuscipes* Curt. (Coleoptera Staphylinidae). Industrie Lito-Tipografiche Mario Ponzio, Pavia. 1.

Pavan, M., and Bo, G. (1952). Ricerche sulla differenziabilita, natura e attivita del principio di *Paederus fuscipes Curt. (Col. Staph.) Mem. Soc. Ent. It.* **31,** 67.

Pavan, M., and Trave, R. (1958). Ants. IV. The venom of *Tapinoma nigerrimum. Insectes Sociaux* , 299.

Phisalix, M. (1922). "Animaux venimeux et venins." Masson, Paris.

Pinder, A. R., and Staddon, B. W. (1965a). Trans-4-oxohex-2-enal in the odoriferous secretion of *Sigara falleni* (Fieb.) (Hemiptera:Heteroptera). *Nature* **205,** 106.

Pinder, A. R., and Staddon, B. W. (1965b). The odoriferous secretion of the water bug, *Sigara falleni* (Fieb.). *J. Chem. Soc.* p.2955.

Poulton, F. B. (1888). The secretion of pure aqueous formic acid by lepidopterous larvae for the purpose of defense. *Brit. Assoc. Advan. Sci., 57th, Manchester (London), 1887,* **765.**

Quilico, A., Piozzi, F., and Pavan, M. (1957a). Ricerche chimiche sui Formicidae: sostanze prodotte dal *Lasius (Chthonolasius) umbratus* Nyl. *Rend. Ist Lombardo Sci.* **91,** 271.

Quilico, A., Grunanger, P., and Piozzi, F. (1957b). Synthesis of tetrahydro- and perhydro-dendrolasin. *Tetrahedron* **1,** 186.

Rawat, B. L. (1939). Notes on the anatomy of *Naucoris cimicoides* L. (Hemiptera:Heteroptera). *Zool. Jahrb.* **65,** 535.

Regnier, F. E., and Law, J. H. (1968). Insect pheromones. *J. Lipid Res.* **9,** 541.

Regnier, F. E., and Wilson, E. O. (1968). The alarm-defense system of the ant *Acanthomyops claviger. J. Insect Physiol.* **14,** 955.

Reichstein, T. (1967). Cardenolide (herzwirksame Glykoside) als Abwehrstoffe bei Insekten. *Naturwissenschaften* **20,** 499.

Reichstein, T., von Euw, J., Parson, J. A., and Rothschild, M. (1968). Heart-poisons in the monarch butterfly. *Science* **161,** 861.

Remold, H. (1963). Scent-glands of land bugs, their physiology and biological function. *Nature* **198,** 764.

Richards, J. H., and Hendrickson, J. B. (1964). "The Biosynthesis of Steroids, Terpenes, and Acetogenins." Benjamin, New York.

Roth, L. M. (1961). A study of the odoriferous glands of *Scaptocoris divergens* Froeschner (Hemiptera:Cydnidae). *Ann. Entomol. Soc. Am.* **54,** 900.

Roth, L. M., and Eisner, T. (1962). Chemical defenses of arthropods. *Ann. Rev. Entomol.* **1,** 107.

Roth, L. M., and Stay, B. (1958). The occurrence of para-quinones in some arthropods, with emphasis on the quinone-secreting tracheal glands of *Diploptera punctata* (Blattaria). *J. Insect Physiol.* **1,** 305.

Roth, L. M., and Willis, E. R. (1960). The biotic associations of cockroaches. *Smithsonian Inst. Misc. Collections* **141,** 1.

Roth, L. M., Niegisch, W. D., and Stahl, W. H. (1956). Occurrence of 2-hexenal in the cockroach *Eurycotis floridana. Science* **123,** 60.

Rothschild, M., Reichstein, T., Aplin, R. T., and Benn, M. H. (1968). Some poisonous Lepidoptera. *Entomol. Soc. Conversazione* **6,** 3.

Rudolph, W. (1916). Cantharidin. *Arch. Pharm.* **254,** 423.

Schildknecht, H. (1957). Zur chemie des bombardierkäfers. *Angew. Chem.* **69,** 62.

Schildknecht, H. (1959). Uber das flüchtige Sekret vom gemeinen Mehlkäfer, II. Mitteilung über Insekten-abwehrstoffe. *Angew. Chem.* **71,** 524.

Schildknecht, H. (1968). Das Arsenal der Schwimmkäfer Sexualhormone und "Antibiotica." *Nach. Chem. Tech.* **18,** 311.

Schildknecht, H., and Holoubek, K. (1961). Die Bombardierkäfer und ihre Explosionschemie, V. Mitteilung über Insekten abwehrstoffe. *Angew. Chem.* **73,** 1.

Schildknecht, H., and Hotz, D. (1967). Identifizerung der Neben steroide des Prothorakabwehrdrüsen systems des Gelbrandkäfers *Dytiscus marginalis. Angew. Chem.* **79,** 902.

Schildknecht, H., and Kornig, W. (1968). Protective material from the prothoracic protective gland of a Mexican *Cybister* species. *Angew. Chem* **7**, 62.

Schildknecht, H., and Kramer, H. (1962). Zum Nachweis von Hydrochinonen neben Chinonen in der Abwehrblasen von Arthropoden. XV. Mitteilung über Insektenabwehrstoffe. *Z Naturforsch.* **17b**, 701.

Schildknecht, H., and Maschwitz, U. (1966). A vertebrate hormone as the defensive substance of the water beetle *Dytiscus marginalis. Angew. Chem* **5**, 421.

Schildknecht, H., and Wenneis, W. F. (1966). Uber Arthropoden-(Insekten) Abwehrstoffe XX. Strukturalklarung des Glomerins. *Z Naturforsch.* **21b**, 552.

Schildknecht, H., and Wenneis, W. F. (1967). Uber Arthropoden-Abwehrstoffe XXV: (1) Anthranilsaure als Precursor der Arthropoden-Alkaloide Glomerin und Homoglomerin. *Tetrahedron Letterss*, p.1815.

Schildknecht, H., and Weis, K. H. (1960). Uber die Tenebrioniden-Chinone bei lebendem und totem Untersuchungsmaterial, VII. Mitteilung uber Insekten-abwehrstoffe. *Z Naturforsch.* **15b**, 757.

Schildknecht, H., and Weis, K. H. (1961a). Chinon als aktives Prinzip der Abwehrstoffe von Diplopoden. *Z Naturforsch.* **16b**, 810.

Schildknecht, H., and Weis, K. H. (1961b). Die chemische Natur des Wehrsekretes von *Pseudophonus pubescens* und *Ps. griseus.* VIII. Mitteilung über Insektenabwehrstoffe. *Z Naturforsch.* **16b**, 361.

Schildknecht, H., and Weis, K. H. (1962). Die Abwehrstoffe einiger Carabiden, insebesondere von *Abax ater.* XII. Mitteilung über Insektenabwehrstoffe. *Z Naturforsch.* **17b**, 439.

Schildknecht, H., Weis, K. H., and Vetter, H. (1962). α-β-Ungesättigte Aldehyde als Inhaltestoffe des Stinkblasender Blattwanze *Dolycoris baccarum* L. *Z Naturforsch.* **17b**, 350.

Schildknecht, H., Holoubek, K., Weis, K. H., and Kramer, H. (1964). Defensive substances of the arthropods, their isolation and identification. *Angew. Chem* **3**, 73.

Schildknecht, H., Wenneis, W. F., Weis, K. H., and Maschwitz, U. (1966). Glomerin, ein neues Arthropoden-Alkaloid. *Z Naturforsch.* **21b**, 121.

Schildknecht, H., Birringer, H., and Maschwitz, U. (1967a). Testosterone as a protective agent of the water beetle, *Ilybius. Angew. Chem* **6**, 558.

Schildknecht, H., Siewerdt, R., and Maschwitz, U. (1967b). Uber Arthropodenabwehrstoffe, XXIII. Cybisteron, ein neues Arthropoden-Steroid. *Ann. Chem* **703**, 182.

Schildknecht, H., Hotz, D., and Maschwitz, U. (1967c). Uber Arthropoden-Abwehrstoffe XXVII. Die C_{21}-Steroide der Prothorakabwehrdrusen von *Acilius sulcatus.* *Z Naturforsch.* **22b**, 938.

Schildknecht, H., Maschwitz, U., and Wenneis, W. F. (1967d). Uber Arthropoden-Abwehrstoffe XXIV. Neue Stoffe aus dem Wehrsekret der Diplopodengattung *Glomeris. Naturwissenschaften* **54**, 196.

Schildknecht, H., Maschwitz, U., and Winkler, H. (1968a). Zur Evolution der Carabiden-Wehrdrusensekrete. Uber Arthropodenabwehrstoffe XXXII. *Naturwissenschaften* **55**, 112.

Schildknecht, H., Maschwitz, U. and Winkler, H. (1968b). Uber Arthropoden-abwehrstoffe. XXXI. Vergleichend chemische Untersuchungen der Inhaltsstoffe der Pygidialwehrblasen von Carabiden. *Z Naturforsch.* **23b**, 637.

Schildknecht, H., Winkler, H., Krauss, D., and Maschwitz, U. (1968c). Uber Arthropoden-abwehrstoffe XVIII. Uber das Abwehrsekret von *Idiochroma dorsalis. Z. Naturforsch.* **23b**, 46.

Schildknecht, H., Maschwitz, U. and Krauss, D. (1968d). Blausaure in Wehrsekret des Erdlaufers *Pachymerium ferrugineum. Naturwissenschaften* **55**, 230.

Staddon, B. W., and Weatherston, J. (1967). Constituents of the stink glands of *Ilyocoris cimicoides* (Heteroptera Naucoridae). *Tetrahedron Letters* p.4567.

Stork, G., Van Tamelen, E. E., Friedman, L. J., and Burgstahler, A. W. (1953). Cantharidin. A stereospecific total synthesis. *J. Am. Chem. Soc.* **75**, 384.

Trave, R., and Pavan, M. (1956). Insect poisons. Extracts of the ant *Tapinoma nigerrimum. Chim. Ind. (Milan)* **38**, 1015.

Trave, R., Garanti, L., and Pavan, M. (1959). Ricerche sulla natura chimica del veleno del miriapode *Archiulus (Schizophyllum) sabulosus* L. *Chim. Ind. (Milan)* **41**, 19.

Tschinkel, W. R. (1969). Phenols and quinones from the defensive secretions of the tenebrionid beetle, *Zophobas rugipes. J. Insect Physiol.* **15**, 191.

von Euw, J., Fishelson, L., Parsons, J. A., Reichstein, T., and Rothschild, M. (1967). Cardenolides (heart poisons) in a grasshopper feeding on milkweeds. *Nature* **214**, 35.

Wain, R. L. (1943). The secretion of salicylaldehyde by the larvae of the brassy willow beetle (*Phyllodecta vittelinae* L.). *Ann. Rep. Agr. Hort. Res. Sta., Long Ashton, Bristol* p.108.

Walter, W. G., and Cole, J. F. (1967). Isolation of cantharidin from *Epicauta pestifera. J. Pharm. Sci.* **56, 174.**

Waterhouse, D. F., Forss, D. A. and Hackman, R. H. (1961). Characteristic odour components of the scent of stink bugs. *J. Insect Physiol.* **6**, 113.

Waterhouse, D. F., and Gilby, A. R. (1964). The adult scent glands and scent of nine bugs of the superfamily Coreidea. *J. Insect Physiol.* **10**, 977.

Waterhouse, D. F., and Wallbank, B. E. (1967). 2 - Methylene butanal and related compounds in the defensive scent of *Platyzosteria* cockroaches (Blattidae: Polyzosteri-inae). *J. Insect Physiol.* **13**, 1657.

Weatherston, J. (1967). The chemistry of arthropod defensive substances. *Quart. Rev. (London)* **21**, 287.

Weatherston, J., and Percy, J. E. (1969). Studies of physiologically active arthropod secretions. III. Chemical, morphological, and histological studies of the defense mechanism of *Uroblaniulus canadensis* (Say) (Diplopoda: Julida).*Can. J. Zool.* **47,**1389.

Wheeler, J. W., Meinwald, J. Hurst, J. J. and Eisner, T. (1964). Trans-2-dodecenal and 2-methyl-1,4-quinone produced by a millipede. *Science* **144**, 540.

Wheeler, W. M, (1890). Hydrocyanic acid secreted by *Polydesmus virginiensis* Drury. *Psyche* **5**, 422.

Wilson, E. O. (1965). Chemical communication in the social insects. *Science* **149**, 1064.

Wilson, E. O., and Bossert, W. H. (1963). Chemical communication among animals. *Recent Progr. Hormone Res.* **19**, 673.

Woodward, R. B., and Loftfield, R. B. (1941). Structure of cantharidin and the synthesis of desoxycantharidin. *J. Am. Chem. Soc.* **63**, 3167.

Yamamoto, I., and Tsuyuki, T. (1964). Odorous principles from stink bugs. *I.U.P.A.C. Intern. Symp. Chem. Nat. Prod. Kyoto* 133.

CURRENT USAGE AND SOME RECENT DEVELOPMENTS WITH INSECT ATTRACTANTS AND REPELLENTS IN THE USDA

Morton Beroza

I. INTRODUCTION

Several topics and investigations within the scope of "chemicals affecting insect behavior" that have been covered only briefly or not at all in the previous chapters remain for consideration. In this connection, the current practice in programs within the USDA relating to insect attractants and repellents will be discussed, and some recent developments will be mentioned. Comment will also be made on the use of attractants and repellents in a general way to present a coherent picture. Some of the work described is aimed at solving problems of practical importance; other problems are of academic interest at the moment, but we hope that some of the findings may lead to useful or practical developments. No attempt is made to review the literature; instead, recent pertinent reviews are cited for those who wish to obtain additional information.

II. ATTRACTANTS

Attractants play a dominant role in many vital aspects of insect behavior (Jacobson, 1965, 1966; Jacobson and Beroza, 1964; Shorey and Gaston, 1967). They govern the activity of insects in seeking food, the opposite sex, or a place to lay their eggs. The action of some of these chemicals—or more specifically their odors—is powerful enough in terms of infinitesimal quantities needed and intensity of response to make them valuable for combating insect pests. In effect, we wish to utilize the survival mechanisms of insects in order to destroy or control them.

A. FINDING SYNTHETIC ATTRACTANTS

One of the objectives within the USDA program is to find synthetic chemical attractants or lures for economically important insect pests. The approach here differs from the isolation route in which an attempt is made to isolate, identify, and synthesize the active ingredient(s) of a natural attractant. Instead, a large number of chemicals are tested against a given insect species to find a lead—i.e., a weak attractant—and then related compounds are synthesized in an effort to secure a more potent attractant. Figure 1 shows a bioassay chamber or olfactometer. Chemicals in traps are exposed to free-flying fruit flies in this chamber to test their attraction.

Over the years, our group has accumulated or synthesized many chemicals which we can test in the initial phases of this screening (Beroza and Green, 1963a,b). Obviously, the greater the number of chemicals tested and the greater their variety, the better are the chances of finding a potent attractant. When a lead is found, the chemist can use his ingenuity and chemical virtuosity in synthesizing related compounds.

Table I lists some of the chemical attractants found by this screening route.

Fɪɢ. 1. Chemicals are tested for fruit fly attraction in glass traps within an 8-foot cubical enclosure stocked with the insects.

B. USE IN DETECTION OF INSECT INFESTATIONS

At present the most widespread use of synthetic attractants is in the trapping of insects for detecting infestations or for determining insect populations. This information is used to direct control or eradication measures and thereby prevent the spread or the flare-up of infestations of damaging insect pests. The first three compounds shown in Table I are being used to detect three of the world's most destructive insect pests, the Mediterranean fruit fly, the melon fly, and the oriental fruit fly. Only the males of the species are attracted by these chemicals.

The three insect pests are present in Hawaii but are not now present on the mainland of the United States. In 1956 the Mediterranean fruit fly was found to infest about a million acres within the State of Florida. With the help of attractant-baited traps to pinpoint the location of the insects, the government was able to eradicate the infestation in 1957 with insecticides, but the cost was steep—about $11,000,000 (Steiner *et al.*, 1961). The citrus industry in Florida

TABLE I
Attractants Found by Volume Screening of Chemicals

Formula	Common or chemical name	Species attracted	References
(structure) Trimedlure formula with Cl, S, CH$_3$, COOC—CH$_3$, CH$_3$	Trimedlure	Mediterranean fruit fly *Ceratitis capitata* (Weid.)	Beroza *et al.* (1961)
CH$_3$C—O—⟨ring⟩—CH$_2$CH$_2$CCH$_3$ (diketone)	Cue-lure	Melon fly *Dacus cucurbitae* Coq.	Alexander *et al.* (1962)
CH$_3$O, CH$_3$O—⟨ring⟩—CH$_2$CH=CH$_2$	Methyl eugenol	Oriental fruit fly *Dacus dorsalis* (Hendel)	Steiner (1952)
CH$_3$CH=CH—CH=CHCH$_2$OCCH$_2$CH$_2$CH$_3$	2,4-Hexadienyl butyrate	Yellow jackets *Vespula* spp.	Davis *et al.* (1967,1968)
CH$_3$(CH$_2$)$_5$CH$_2$OCCH$_2$CH$_2$CH$_3$	Heptyl butyrate	Yellow jackets *Vespula* spp.	Davis *et al.* (1969)
(benzodioxane structure) O, H$_2$, H, COOCH$_2$CH$_2$CH$_3$	Amlure	European chafer *Amphimallon majalis* (Raz.)	McGovern *et al.* (1969b)
(ring) S, CH$_2$CH$_2$COCH$_3$, O and CH$_3$O, HO—⟨ring⟩—CH$_2$CH=CH$_2$	Methyl cyclohexane-propionate and eugenol	Japanese beetle *Popillia japonica* (Newman)	McGovern *et al.* (1969a)
H$_3$C, H$_3$C, CHCH$_2$CHCHCOOCH$_2$CH$_3$, H$_3$C—C—CH$_3$	Ethyl dihydrochrysanthe-mumate (ethyl 3-isobutyl-2,2-dimethylcyclopropane-carboxylate)	Rhinoceros beetle *Oryctes rhinoceros* (L.)	Barber *et al.* (1970)

estimated it would cost $20,000,000 annually to live with the pest (Reagan, 1966).

The great amount of international trade and traffic made it inevitable that accidental imports of this kind would occur again, and they have. With attractant-baited traps deployed about our ports of entry, however, new infestations were rapidly detected and eradicated before they could spread (Reagan, 1966). USDA officials estimated that between 1958 and 1964 this early warning detection system saved the government at least $9,000,000 in potential eradication costs through rapid detection and eradication of the reinfestations (Beroza, 1966).

The trap used in this program is shown in Fig. 2. It does triple-duty by holding the three lures just cited. When responding to the lure, the target insects are overcome by a volatile insecticide within the trap and thereby betray their presence. An idea of the potency and specificity of these lures can be gained from the fact that a trap baited with methyl eugenol *only* is often as much as one-third full of oriental fruit flies after a 1-week exposure in Hawaii, and no other species of insect is found in the trap.

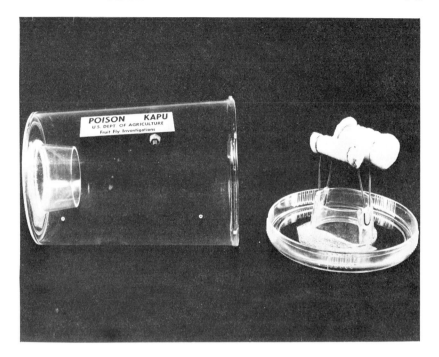

FIG. 2. Plastic trap (disassembled) holds attractants for Mediterranean fruit fly, oriental fruit fly, and melon fly on cotton roll dispensers.

C. NEWLY FOUND ATTRACTANTS

The remaining attractants of Table I were found since the listing of synthetic attractants by Beroza and Green (1963a). Unlike the first three attractants of Table I, which attract only males, these chemicals attract both males and females.

Yellow jacket wasps have pestered many of us; they are also a menace to beekeepers and to farmers who grow certain crops. In preliminary tests 2,4-hexadienyl butyrate and heptyl butyrate (Davis *et al.*, 1967–1969) have already proved useful in alleviating these difficulties by monitoring and trapping the yellow jackets in a given area.

Amlure (McGovern *et al.*, 1970b) catches about three times as many European chafers as butyl sorbate, which was formerly considered the best attractant for this insect (Tashiro *et al.*, 1964). Presently the chafer is restricted to several small localities in the northeastern part of the country. Amlure is helping our action program people in their efforts to prevent further spread of the European chafer.

The Japanese beetle now infests almost the entire eastern part of our country. The attractant mixture of methyl cyclohexanepropionate and eugenol (McGovern *et al.,* 1970a) is about three times as effective as the phenethyl butyrate–eugenol combination which was previously the best lure used (Schwartz *et al.,* 1966).

The economy of many Pacific Islands depends on copra, the meat of the coconut. The rhinoceros beetle has recently appeared on many Pacific Islands and is causing considerable damage to coconut trees. In Western Samoa, ethyl dihydrochrysanthemumate was shown to be an excellent attractant for the beetle (Barber *et al.,* 1970). The rhinoceros beetle is about 2 inches long. It flies to the top of a coconut tree and consumes the growing tip, thus killing the tree. This is a very serious problem, and we hope the attractant will be useful in efforts to control this insect.

D. STRUCTURE OF ATTRACTANTS

The synthetic attractants listed include a variety of structures— esters, ethers, ketones, phenols, unsaturates, aromatics, carbocyclics, and heterocyclics. If anything, the synthetic attractants may be characterized by their diversity rather than by any similarity. Inasmuch as the good attractants are usually highly specific for a given species, one may conclude that such diversity of structure is to be expected. One may wish to conclude further that the structures of insect attractants are simply unpredictable.

We do not know why the various insects respond to these synthetic lures. The lures could represent a source of food to these insects, or those that attract only the males of a species could have an odor similar to that of the sex attractant of that species; e.g., Fletcher (1968) has suggested that there may be a relationship between synthetic lures used in the control of certain fruit flies and the male pheromone of the species.

Consider trimedlure, the Mediterranean fruit fly attractant. It is composed of the four isomers shown in Fig. 3 (McGovern and Beroza, 1966). No such compounds are known to occur in nature, yet this material is a powerful attractant. We have carefully separated and determined the structure of the individual isomers, and tested the attraction of each. One of the isomers does not attract the Mediterranean fruit fly at all (McGovern *et al.,* 1966), yet the only difference in

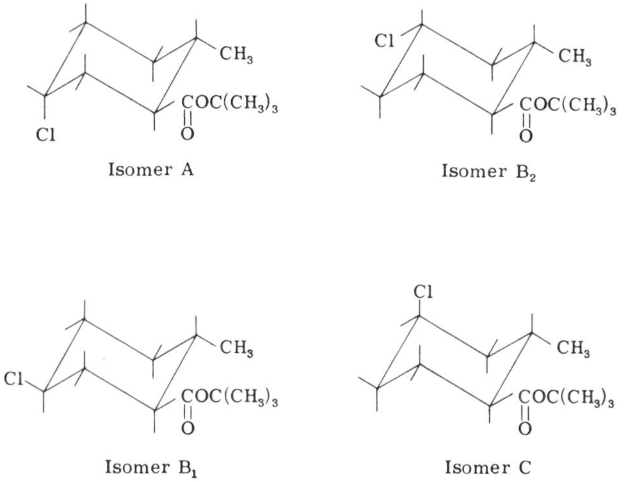

FIG. 3.   Isomers of trimedlure.

the four isomers is the position of the chlorine atom. This small difference in molecular architecture illustrates how similar compounds can be and still possess attractive properties that are quite different.

E. ATTRACTANT TECHNOLOGY

One of the most discouraging aspects of working with attractants is the great amount of patience and expense involved in finding a truly effective chemical. Frequently, thousands of compounds have had to be procured or synthesized and then tested. Yet, if we consider that an attractant, once found and properly used, is available for all future time, or that it may provide a means of safe pest control, the search is decidedly worthwhile. Finding an attractant, however, is only part of the job. Each insect species has its own idiosyncrasies. The entomologist has to design a trap that will give maximum catches (Holbrook *et al.,* 1960). He has to check its color, the size and type of bait dispenser, the amount of chemical, size of the openings, and the trapping technique (which may utilize a volatile insecticide, a sticky substance, a mechanical baffle, or a detergent liquid in the trap). Sometimes there are unusual interferences. When ants were found to be stealing trapped fruit flies, an ant repellent had to be added to the trap. Birds also began to steal the trapped flies, and a screen covering half of the trap openings had to be installed to exclude them (see Fig. 2). Even a well-designed trap will not catch the flies if improperly placed. Traps placed in hollows or in dense growth yield poor catches because the restricted air movement does not allow the vapors of the attractant chemical to be carried away. The height of a trap can also affect catches.

There are other subtleties too. An attractant found effective in a laboratory bioassay sometimes does not attract in the field. Jacobson found that the natural attractant of the pink bollworm moth did not attract in the field after it was isolated through the use of a laboratory bioassay. Fortunately, the addition of another substance produced the desired result (Jones and Jacobson, 1968; see chapter by Jacobson *et al.* in this volume). This difficulty is also encountered in the screening of synthetics in a laboratory bioassay. Ultimately, any attractant that is found must prove itself in the field in direct competition with the many natural odors, colors, and other stimuli present under the actual conditions of use.

More and more we are encountering this need to use combinations of chemicals. Silverstein (in his chapter in this volume) reported that several ingredients were needed to attract the western pine beetle. Tumlinson and his co-workers showed the natural boll weevil attractant to be a combination of four chemicals rather than a pure compound (see their chapter in this volume). This situation also holds with the pure synthetics. Thirty years ago, a geraniol–eugenol combination was shown to be a better lure for the Japanese beetle than either individual ingredient used separately (Fleming *et al.,* 1940). At present our best Japanese beetle lure is still a mixture of two chemicals.

The dispensing of these lure mixtures presents problems: What is the optimum proportion of each chemical? How can this proportion be maintained when the chemicals have different volatilities? We are facing the latter problem with the new Japanese beetle attractant mixture because the methyl cyclohexanepropionate is much more volatile than eugenol, the other ingredient. Our experiments thus far indicate that despite their different volatilities the chemicals can be dispensed in the desired proportion from a reservoir fitted with a wick (McGovern and Beroza, 1970). Thus, the composition of the chemicals in the reservoir remains reasonably constant as the ingredients volatil-ize from the wick. Undoubtedly, the proportion of the less volatile chemical on the exposed portion of the wick becomes greater than that in the reservoir, and a steady-state situation is set up until the reservoir contents are exhausted.

We have heard of the masking of chemical odors. We do not know enough about this effect, which is encountered with both natural (Beroza, 1967; Jacobson, 1969; Jacobson and Smalls, 1966, 1967) and synthetic (Waters and Jacobson, 1965) attractants. There is evidence, however, that compounds closely related in chemical structure to an attractant can mask the attractant. We can speculate that the olfactory receptor sites are blocked by the masking compound.

Vapor concentration is important. The ideal lure will attract at high or low concentrations, but few chemicals act this way. Instead they repel at high concentrations and cannot be detected at low concentrations.

Let us also note that attractant vapors are heavier than air and will tend to fall before being completely dispersed. Many insects have accommodated their behavior to this fact and pursue odors while flying just above the ground, the level likely to have maximum attractant concentration on nonwindy days.

A frequently asked question is how much attractant is used by the United States Government, the major purchaser of attractants. *For detection purposes*, the amount is not great—usually from 200 to 2000 pounds of an individual attractant a year, depending on the area being surveyed, the volatility of the chemical, duration of trapping, and other factors.

F. USE IN CONTROL

If an attractant is to be used for direct control, e.g., by combining it with a toxicant, large quantities of the chemical will be needed. Inexpensive protein hydrolyzates that attract insects generally rather than specifically have been combined with insecticides to minimize the use of insecticide and increase the efficacy of the insecticide. Complete coverage of an area is then unnecessary because the insects are attracted to the poisoned protein hydrolyzate, consume it, and die. A malathion-protein hydrolyzate spray of this type was used in the eradication of the Mediterranean fruit fly in Florida in 1957 (Steiner *et al.,* 1961).

In at least one instance a specific lure has been combined with insecticide for direct control. Methyl eugenol was used to eradicate the oriental fruit fly from Rota, a small island in the Pacific Ocean (Steiner *et al.,* 1965). Department entomologists dropped from airplanes small absorbent cane-fiber squares containing a combination of the lure and naled, a fast-acting insecticide. Drops were made about every 2 weeks, and the insect was eradicated in several months even though the lure attracts only males. The advantage of this approach is that only the offending insect was attacked. Wildlife was not affected, and insecticides were not spread needlessly.

G. ATTRACTANTS AND THE OLFACTORY PROCESS

In reading the literature on attractants, one is struck by the frequent reference to our lack of knowledge of the olfactory process (Schneider, 1969). Insects are such readily available creatures that they would appear to be most useful as experimental animals on which to develop our knowledge in this area, especially since some very powerful attractants that can be used as tools are available for this purpose.

One such investigation recently reported (Doolittle *et al.,* 1968) was aimed at determining the validity of certain theories of olfaction. One theory states that stereochemical features of a molecule or its shape can determine its odor (Amoore *et al.,* 1964). Another, the Dyson–Wright theory, states that vibrations of the molecule in the far-infrared region (500 to 50 cm^{-1}). trigger the olfactory response (Wright, 1968). We used cue-lure as the attractant and systematically replaced the hydrogen atoms with deuterium in various parts of the molecule, as well as replacing the hydrogen atoms almost completely. The positions of deuteration were established by NMR spectra of the preparations. The heavier deuterium atoms can be expected to shift the infrared absorption maxima of the original cue-lure to lower frequencies; such shifts would be expected to change the odor, hence the attractiveness, according to the Dyson–Wright theory. These shifts did occur following deuteration as shown in Fig. 4, but despite extensive testing involving more than 2600 counts of responding insects there was no difference in attraction between cue-lure and any of the deuterated compounds. We had to conclude that our data did not support the infrared absorption theory for odor.

The proponents of molecular shape as the determinant of odor may find these data useful, since deuteration of the molecule does not change its shape; deuteration also did not change the odor, if we accept the response of the insects as a test of such a change.

H. ATTRACTION OF MOSQUITOES

Another recent study also sheds some light on the olfactory process. For years scientists the world over have been unable to

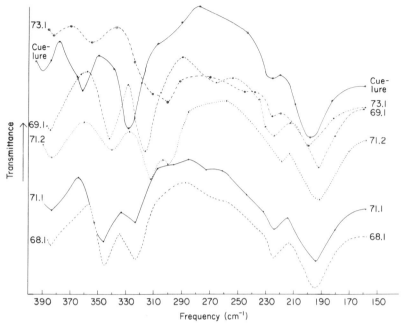

FIG. 4. Far-infrared spectra of cue-lure and analogs deuterated in various parts of the cue-lure molecule (Doolittle *et al.*, 1968).

determine how a female mosquito detects a human, which it probes for a blood meal. Recently it was shown that L-lactic acid plus CO_2 attracted female yellow-fever mosquitoes (Acree *et al.*, 1968). (L-Lactic acid is about five times as attractive as its D-isomer.) The insects did not respond to either chemical singly, but only to the combination. The vapors of the two chemicals, however, are not the only attractants. Heat, moisture, and even sound may play a role. Thus, physical agents can be implicated in the attraction process. Incidentally, efforts are now in progress to use lactic acid and CO_2 for yellow-fever mosquito detection since the worldwide eradication of this insect is being attempted.

III. REPELLENTS

Repellents have tremendous practical importance (Garson and Winnike, 1968; Painter, 1967). Thus, a considerable effort by many scientists in a program to protect our Armed Forces from insect-borne diseases has been expended in finding chemicals to protect people directly from a variety of dangerous or pestiferous insects. In parts of the world known to be infested with malaria-carrying mosquitoes, typhus-carrying body lice, and other arthropod vectors of disease, the use of repellents is virtually essential.

A. REQUIREMENTS FOR A GOOD REPELLENT

USDA scientists have collected, synthesized, and tested more than 20,000 chemicals as repellents against a variety of arthropods (Entomology Research Division, 1967; King, 1954). Although many of the compounds were effective, few could be used. Compounds to be placed on human skin or clothing must be nontoxic, nonirritating, nonallergenic, harmless to wearing apparel, inoffensive in odor, long-lasting, stable to sunlight, and effective against as broad a spectrum of insect pests as possible. The skin repellents should withstand sweating and resist loss by wiping; clothing repellents should resist leaching by rain and laundering. Requirements are so stringent that of the chemicals tested only about two in a thousand were safe enough for use on skin. The emphasis on toxicity is justified when one considers that repellents are often used over a large portion of the body for extensive periods of time, especially in tropical areas. As part of this evaluation, absorption, retention, metabolic changes, mode of excretion, damage to delicate tissue (e.g., the eye), inhalation vs. absorption toxicity, residues, and long-term (chronic) effects on various organs are determined.

At the moment there appears to be little hope of repelling pestiferous insects economically from a large area of operation by a chemical, and the only practical resort is to apply a chemical to the skin or clothing of the individual to keep the arthropod at a distance. The most widely used repellent for application to skin today is deet (*N,N*-diethyl-*m*-toluamide), a product of the USDA research program (Gilbert *et al.,* 1957; McCabe *et al.,* 1954). It is effective against fleas, mosquitoes, chiggers, ticks, deer flies, sand flies, biting gnats, and land leeches. When applied to skin, it protects for 3 to 8 hours against a dozen insect species in tests conducted in areas from Panama to Alaska.

B. STRUCTURE OF REPELLENTS

What kinds of chemicals are insect repellents? There is quite a variety (Garson and Winnike, 1968). Deet is an amide, but imides, alcohols, esters, ethers, and such polyfunctional compounds as 1,3-diols, amide esters, hydroxy esters, and diesters are among the compounds found to be good repellents. Attempts to relate repellency to specific chemical or physical features of a compound have not been sufficiently informative to allow the prediction of repellency with any degree of accuracy.

C. CLOTHING REPELLENTS

Deet does not persist on clothing exposed to water rinsing or laundering (Gertler *et al.,* 1962). Several clothing repellents have been developed which do persist on clothing. One of the best is the so-called "M-1960" formulation (Acree and Beroza, 1962). Is is an emulsifiable concentrate containing by weight: 30% benzyl benzoate (protects against mites), 30% *N*-butylacetanilide (protects against ticks), 30% 2-butyl-2-ethyl-1,3-propanediol (protects against mosquitoes), and 10% Tween 80 (emulsifier). M-1960 is applied at 2 gm per square foot and is effective for several weeks. Like deet it also repels land leeches. This formulation illustrates the use of several chemicals to repel more species than is possible with any one of the chemicals alone.

D. SOME OTHER TYPES OF REPELLENTS

Chemicals are also used to protect domestic animals from insects, but these are usually toxicants. Pyrethrins, malathion, methoxychlor, and other insecticides have been applied, generally by spraying, dipping, dusting, or by using a back rubber. A good inexpensive insect repellent for use on livestock would boost milk and meat production and find widespread use. An animal plagued by insects expends much energy, energy that is normally used for growth. Thus far, a satisfactory chemical for this purpose has not been found.

Attempts to find a safe systemic repellent (one that is taken internally) have been mentioned, but none have yet been forthcoming (Sherman, 1966; Smith, 1966).

Nor has anyone developed an insect repellent for use on agricultural products. A volatile repellent does not last long enough when the duration of protection required is considered, and frequent spraying of a repellent over the long growing season is not economically feasible. Nonvolatile materials are really feeding deterrents or antifeedants (Wright, 1967). They are effective where deposited, but new tissue not covered by the spray is readily attacked by insects so that the growing plant is not protected.

FIG. 5. Space repellent field test. Effectiveness of a chemical is measured by the number of days mosquitoes are unable to pass through the chemically treated netting (¼-inch square holes) and reach the subject.

E. SPACE REPELLENTS

For a number of years now entomologists have been experimenting with space repellents for mosquitoes (Gouck *et al.,* 1967; McGovern *et al.,* 1967). Space repellents are so named because they exert repellency a small distance from the site of application. These chemicals are applied to nets with ¼-inch square openings and give excellent protection, even though mosquitoes can pass the untreated nets readily. The advantage of using space repellents on nets is that air-flow through these nets is less impeded than with the usual nets with very small holes, and the comfort of the individual is greatly enhanced, especially in tropical countries. Such a net used to cover the entrance to a tent can protect those inside and still allow air to move in and out of the tent. Space repellents should be easier to find than skin or clothing repellents because they do not contact the skin, and the pharmacological requirements for clearance of these materials should be less stringent. The repellents on treated nets have also been found to last for months. Compare this with skin repellents which usually do not last 1 day.

Figure 5 shows a chemically treated netting being tested in the field. The subject counts the flies that penetrate the netting.

IV. APPRAISAL OF CHEMICALS FOR CONTROLLING
INSECT BEHAVIOR

In considering the presentations of this Symposium, some may wonder whether attractants, repellents, or some of the hormonelike materials will supplant the use of insecticides. This hope may stem from the statements we read from time to time on the deleterious effects of pesticides on our ecological systems (Edwards, 1969; Hickey and Anderson, 1968; West, 1966; Wurster, 1968; Wurster and Wingate, 1968) or from the concern expressed for human safety (Lofroth, 1968). At least in the forseeable future, it appears that insecticides will continue to be our mainstay in combating insects. Insecticides are a practical necessity for providing the food and fiber needed by ourselves and much of the world, and the precautions taken these days to assure the safe use of insecticides are most thorough (Knipling, 1966). Attractants, however, will make it possible to control the insect population using less insecticide by showing where and when insecticides should be applied and by directing chemical attack selectively against insect pests in order to avoid harm to fish, birds, or other wildlife. Attractants, by making possible the monitoring of insect populations, will be invaluable for conducting eradication programs efficiently where such possibilities exist. The unheralded network of attractant-baited traps about our ports of entry will continue to prevent the accidental importation of certain foreign insect pests. As long as these pests are kept out, no pesticide is needed to control them. Thus, insect attractants will not eliminate insecticides, but they will enable us to use insecticides wisely.

The use of insect repellents on humans is today firmly established, and obviously these materials do diminish the need for insecticides in many situations. From the viewpoint of ecology, repellents are desirable control chemicals because they offer protection and comfort and do not disturb any ecosystems. There is little doubt that progress in devising and utilizing repellents will continue. Hopefully, someone will find a good insect repellent for livestock. Defensive secretions (see chapter by Weatherston and Percy in this volume) or behavior-inducing chemicals used by the insects themselves (see chapter by Blum in this volume) may provide a clue that will help us to devise a good repellent.

The trend today is to use all available weapons or tools against insect pests, that is, the so-called integrated approach (Chant, 1966; Hoyt, 1969; Smith and van den Bosch, 1967). There is little doubt that all the tools and methods (or improved ones) discussed in this Symposium will be mustered in our efforts toward minimizing pesticide residues and in assuring the safety of our people, our food supply, our environment values, and our wildlife (Knipling, 1966). In essence then, if pest control operations are considered safe today, attractants, repellents, and the other chemical means of combating insect pests that were cited will make these operations safer.

SUMMARY

Synthetic insect attractants are being used to reduce the amount of insecticide needed to control or eradicate harmful insect species. Means of finding and using attractants, their value, their structures, and their use in deciphering olfactory processes are discussed. In a discussion of insect repellents, the value and requirements of a good repellent, their types, and structures are reviewed briefly. These and other chemicals that control insect behavior are making possible significant reductions in insecticide residues and are setting the stage for new developments in pest control.

REFERENCES

Acree, F., Jr., and Beroza, M. (1962). Quantitative gas chromatography of insect repellent mixture M-1960. *J. Econ. Entomol.* **55**, 469.
Acree, F., Jr., Turner, R. B., Gouck, H. K., Beroza, M., and Smith, N. (1968). L-Lactic acid: a mosquito attractant isolated from humans. *Science* **161**, 1346.
Alexander, B. H., Beroza, M., Oda, T. A., Steiner, L. F., Miyashita, D. H., and Mitchell, W. C. (1962). The development of bait for the male melon fly. *J. Agr. Food Chem.* **10**, 270.
Amoore, J. F., Johnston, J. W., Jr., and Rubin, M. (1964). The stereochemical theory of odor. *Sci. Am.* **210**, (2), 42.
Barber, I., McGovern, T. P., and Beroza, M. (1970). Manuscript in preparation.
Beroza, M. (1966). Future role of natural and synthetic attractants for pest control. . *U. S. Dept. Agr.* **ARS 33-110**, 34.
Beroza, M. (1967). Nonpersistent inhibitor of the gypsy moth sex attractant in extracts of the insect. *J. Econ. Entomol.* **60**, 975.
Beroza, M., and Green, N. (1963a). Synthetic chemicals as insect attractants. *Advan. Chem. Ser.* **41**, 11.
Beroza, M., and Green, N. (1963b). Materials tested as insect attractants. *U. S. Dept. Agr. Agr. Handbook* **239**.
Beroza, M., Green, N., Gertler, S. I., Steiner, L. F., and Miyashita, D. H. (1961). New attractants for the Mediterranean fruit fly. *J. Agr. Food Chem.* **9**, 361.
Chant, D. A. (1966). Integrated control systems. *In* "Scientific Aspects of Pest Control." *Natl. Acad. Sci.-Natl. Res. Council, Publ.* **1402**, 193-218.
Davis, H. G., Eddy, G. W., McGovern, T. P., and Beroza, M. (1967). 2,4-Hexadienyl butyrate and related compounds highly attractive to yellow jackets (*Vespula* spp.). *J. Med. Entomol.* **4**, 275.
Davis, H. G., McGovern, T. P., Eddy, G. W., Nelson, T. E., Bertun, K. M. R., Beroza, M., and Ingangi, J. C. (1968). New chemical attractants for yellow jackets (*Vespula* spp.). *J. Econ. Entomol.* **61**, 459.
Davis, H. G., Eddy, G. W., McGovern, T. P., Beroza, M. (1969). Heptyl butyrate, a new synthetic attractant for yellow jackets. *J. Econ. Entomol.* **62**, 1245.
Doolittle, R. E., Beroza, M., Keiser, I., and Schneider, E. L. (1968). Deuteration of the melon fly attractant, cue-lure, and its effect on olfactory response and infrared absorption. *J. Insect Physiol.* **14**, 1697.
Edwards, C. A. (1969). Soil pollutants and soil animals. *Sci. Am.* **220**, (4) 88.
Entomology Research Division (1967). Materials evaluated as insecticides, repellents, and chemosterilants at Orlando and Gainesville, Fla., 1952-1964. *U. S. Dept. Agr. Agr. Handbook* **340**.

Fleming, W. E., Burgess, E. D., and Maines, W. W. (1940). The use of traps against the Japanese beetle. *U. S. Dept. Agr. Circ.* **594**.

Fletcher, B. S. (1968). Storage and release of a sex pheromone by the Queensland fruit fly, *Dacus tryoni* (Diptera: Trypetidae). *Nature* **219**, 631.

Garson, L. R., and Winnike, M. E. (1968). Relationships between insect repellency and chemical and physical parameters—a review. *J. Med. Entomol.* **5**, 339.

Gertler, S. I., Gouck, H. K., and Gilbert, I. H. (1962). *N*-Alkyl toluamides in cloth as repellents for mosquitoes, ticks, and chiggers. *J. Econ. Entomol.* **55**, 451.

Gilbert, I. H., Gouck, H. K. and Smith, C. N. (1957). New insect repellent. *Soap Chem. Specialities* **33**, 115, 129; **33**, 95, 109.

Gouck, H. K., McGovern, T. P., and Beroza, M. (1967). Chemicals tested as space repellents against yellow-fever mosquitoes. I. Esters. *J. Econ. Entomol.* **60**, 1587.

Hickey, J. J., and Anderson, D. W. (1968). Chlorinated hydrocarbons and eggshell changes in raptorial and fish-eating birds. *Science* **162**, 271.

Holbrook, R. F., Beroza, M., and Burgess, E. D. (1960). Gypsy moth *(Porthetria dispar)* detection with the natural female sex lure. *J. Econ. Entomol.* **53**, 751.

Hoyt, S. C. (1969). Integrated chemical control of insects and biological control of mites on apple in Washington. *J. Econ. Entomol.* **62**, 74.

Jacobson, M. (1965). "Insect Sex Attractants." Wiley, New York.

Jacobson, M. (1966). Chemical insect attractants and repellents. *Ann. Rev. Entomol.* **11**, 403.

Jacobson, M. (1969). Sex pheromone of the pink bollworm moth: biological masking by its geometrical isomer. *Science* **163**, 190.

Jacobson, M., and Beroza, M. (1964). Insect attractants. *Sci. Am.* **211**, (2) 20.

Jacobson, M., and Smalls, L. A. (1966). Masking of the American cockroach sex attractant. *J. Econ. Entomol.* **59**, 414.

Jacobson, M., and Smalls, L. A. (1967). Sex attraction masking in the cynthia moth. *J. Econ. Entomol.* **60**, 296.

Jones, W. A., and Jacobson, M. (1968). Isolation of *N,N*-diethyl-*m*-toluamide (deet) from female pink bollworm moths. *Science* **159**, 99.

King, W. V. (1954). Chemicals evaluated as insecticides and repellents at Orlando, Fla. *U. S. Dept. Agr. Agr. Handbook* **69**.

Knipling, E. F. (1966). New horizons and the outlook for pest control. *In* "Scientific Aspects of Pest Control." *Natl. Acad. Sci.–Natl. Res. Council, Publ.* **1402**, 455-470.

Lofroth, G. (1968). Pesticides and catastrophe. *New Scientist* **40**, 567.

McCabe, E. T., Barthel, W. F., Gertler, S. I., and Hall, S. A. (1954). Insect repellents. III. *N,N*-Diethyltoluamides. *J. Org. Chem.* **19**, 493.

McGovern, T. P., and Beroza, M. (1966). Structure of the four isomers of the insect attractant trimedlure. *J. Org. Chem.* **31**, 1472.

McGovern, T. P. and Beroza, M. (1970). Volatility and compositional changes of Japanese beetle attractant mixtures and means of dispensing sufficient vapor having a constant composition. *J. Econ. Entomol.* submitted.

McGovern, T. P., Beroza, M., and Gouck, H. K. (1967). Chemicals tested as space repellents for yellow-fever mosquitoes. II. Carbanilates, benzamides, aliphatic amides, and imides. *J. Econ. Entomol.* **60**, 1591.

McGovern, T. P., Beroza, M., Ohinata, K., Miyashita, D., and Steiner, L. F. (1966). Volatility and attractiveness to the Mediterranean fruit fly of trimedlure and its isomers, and a comparison of its volatility with that of seven other insect attractants. *J. Econ. Entomol.* **59**, 1450.

McGovern, T. P., Beroza, M., Schwartz, P. H., Hamilton, D. W., Ingangi, J. C., and Ladd, T. L. (1970a). Methyl cyclohexanepropionate and related chemicals as attractants for Japanese beetles. *J. Econ. Entomol.* **63**, 276.

McGovern, T. P., Fiori, B., Beroza, M., and Ingangi, J. C. (1970b). Propyl 1,4-benzodioxan-2-carboxylate, a potent attractant for the European chafer. *J. Econ. Entomol.* **63**, 168.

Painter, R. R. (1967). Repellents. *In* "Pest Control: Biological, Physical, and Selected Chemical Methods" (W. W. Kilgore and R. L. Doutt, eds.), pp. 267-285. Academic Press, New York.

Reagan, E. P. (1966). Preventive pest-control measures. *In* "Scientific Aspects of Pest Control." *Natl. Acad. Sci.–Natl. Res. Council, Publ.* **1402.**, 185-192.

Schneider, D. (1969). Insect olfaction: deciphering system for chemical messages. *Science* **163**, 1031.

Schwartz, P. H., Hamilton, D. W., Jester, C. W., and Townshend, B. G. (1966). Attractants for Japanese beetles tested in the field. *J. Econ. Entomol.* **59**, 1516.

Sherman, J. L., Jr. (1966). Development of a systemic insect repellent. *J. Am. Med. Assoc.* **196**, 256.

Shorey, H. H., and Gaston, L. K. (1967). Pheromones. *In* "Pest Control: Biological, Physical, and Selected Chemical Methods" (W. W. Kilgore and R. L. Doutt, eds.), pp. 241-265. Academic Press, New York.

Smith, C. N. (1966). Personal protection from blood-sucking arthropods. *J. Am. Med. Assoc.* **196**, 236.

Smith, R. F., and van den Bosch, R. (1967). Integrated control. *In* "Pest Control: Biological, Physical, and Selected Chemical Methods" (W. W. Kilgore and R. L. Doutt, eds.), pp. 295-340. Academic Press, New York.

Steiner, L. F. (1952). Methyl eugenol as an attractant for oriental fruit fly. *J. Econ. Entomol.* **45**, 341-348.

Steiner, L. F., Rohwer, G. G., Ayers, E. L., and Christenson, L. D. (1961). The role of attractants in the recent Mediterranean fruit fly eradication program in Florida. *J. Econ. Entomol.* **54**, 30.

Steiner, L. F., Mitchell, W. D., Harris, E. J., Kozuma, T. T., and Fujimoto, M. S. (1965). Oriental fruit fly eradication by male annihilation. *J. Econ. Entomol.* **58**, 961.

Tashiro, H., Gertler, S. I., Beroza, M., and Green, N. (1964). Butyl sorbate as an attractant for the European chafer. *J. Econ. Entomol.* **57**, 230.

Waters, R. M., and Jacobson, M. (1965). Attractiveness of gyplure masked by impurities. *J. Econ. Entomol.* **58**, 370.

West, I. (1966). Biological effects of pesticides in the environment. *Advan. Chem. Ser.* **60**, 38.

Wright, D. P., Jr. (1967). Antifeedants. *In* "Pest Control: Biological, Physical, and Selected Chemical Methods" (W. W. Kilgore and R. L. Doutt, eds.), pp. 287-293. Academic Press, New York.

Wright, R. H. (1968). How animals distinguish odours. *Science J.* **4**, 57.

Wurster, C. F., Jr. (1968). DDT reduces photosynthesis by marine phytoplankton. *Science* **159**, 1474.

Wurster, C. F., Jr., and Wingate, D. B. (1968). DDT residues and declining reproduction in the Bermuda petrel. *Science* **159**, 979.

INDEX

A

Acanthomyops claviger Roger
  defensive secretions of, 99, 104
  response to alarm pheromone, 87
Acids, aliphatic, in defensive secretions,
    117, 118
Aldehydes, aliphatic, in defensive secre-
    tions, 117, 118
Alkaloids in defensive secretions,
    117, 135
Amlure, as attractant for European
    chafer, 148, 149
Antennae, as chemoreceptors, 4, 78, 80
Anthonomus grandis Boheman,
    see Boll weevil
Ants
  dolichoderine, 63, 87, 101-104
  doryline, 80, 82
  formicine, 74-75, 99-101
  myrmicine, 63, 81, 99-101
  Nearctic army ants, interspecies
      attraction to queen odors, 80
  necrophoric behavior, 76
  sex pheromones of, 74-75
  terpenoid compounds in defensive
      secretions of, 98-104
Apheloria corrugata, defensive secre-
    tion of, 114, 136
Apis mellifera, see Honeybee
Apis spp., nonspecificity of pheromones,
    79-80
Aristolochic acid in butterflies, food
    plants as source of, 132-133
Aromatic compounds in defensive secre-
    tions, 112-116
Arthropods, classification of, 96-97
Atta texana (Buckley)
  alarm pheromone of, 77, 78
  trail pheromone of, 65
Attagenus megatoma (Fabricius),
    see Black carpet beetle
Attractants, *see also* Pheromones,
    individual insects

dispensing mixtures of, 152
masking of, 7, 15-16, 24, 153
practical considerations, 151-153
sex attractants
  activator for, 7
  effective range of, 14, 34
  of formicine ants, 74-75
  of honeybee, 72-74, 79-80
  potential uses of, 14-16
synthetic 146-155
  amounts used, 153
  screening for, 146
use in insect control, 153-154
use in insect surveys, 4, 147-8

B

Benzaldehyde
  as aphrodisiac, 69
  in defensive secretions, 69, 114
  as trail pheromone, 67-69, 83
Benzoic acid, in defensive secretions, 114
Bioassay
  of insect attractants 30, 51, 54,
      146, 152
  in following isolation of pheromones, 25
Biosynthesis
  of benzoquinones, 111-112
  of boll weevil attractants, 54-55
  of defensive secretions, 135-136
  of hydrogen cyanide, 114
  of terpenoid defensive substances,
      104-105
Black carpet beetle, megatomic acid as
    attractant for, 22, 33-34
Bombykol, sex pheromone of silkworm,
    10
Boll weevil, 22
  attractants of, 41-59, 152
Brevicomin
  attractant of western pine beetle,
      29-32
  identification and synthesis, 30-32
  isomers, 29

165

Brevicomin *(continued)*
  response of bark beetle predator to, 30
Bumblebees, multiple functions of mandibular gland secretions of, 75
Butyl sorbate, attractant for European chafer, 149

2-heptanone as alarm pheromone of, 87
Cotton boll weevil, *see* Boll weevil
Cue-lure, attractant for melon fly, 148
  effect of deuteration on properties of, 154
Cuminaldehyde, in defensive secretion, 114

C

Cabbage looper
  compounds tested as attractants for, 11-13
  isolation and synthesis of sex pheromone, 7-8
  use of synthetic pheromone for control of, 14-15
Cantharidin, as defensive substance, 132
Carbon dioxide in attraction of mosquitoes, 155
Cardiac glycosides, in danaiid butterflies and grasshoppers, 132-133
Caryophyllene and caryophyllene oxide in boll weevil plant attractant, 42-43
Ceratitis capitata (Wiedemann), *see* Mediterranean fruit fly
Chauliognathus lecontei, dihydromatricaria acid in defensive secretion of, 117
Chemosterilant research with boll weevils, 42
Citral
  as alarm pheromone, 71
  in defensive secretions, 88, 99, 104
  multiple functions of, in *Lestrimelitta limao,* 87-88
  from Nassanoff gland, 73
Citronellal
  as alarm pheromone, 71
  in defensive secretions, 99, 104
Citronellol, in ants, 74, 99
Codling moth, compounds tested as attractants, 11, 13
Coleoptera, attractant pheromones of, 21-40, *see also individual species*
Conomyrma pyramica (Roger),

D

Dacus cucurbitae Coquillet, *see* Melon fly
Dacus dorsalis Hendel, *see* Oriental fruit fly
Deet, 56
  as activator of propylure, 7
  as repellent, 156, 157
Defensive secretions, 88, 95-144
  derived from food plants, 132-133, 135
  role of secondary compounds in, 106
  various forms of, 96
Dendroctonus brevicomis LeConte, *see* Western pine beetle
Dendroctonus frontalis Zimmerman, *see* Southern pine beetle
Dendroctonus ponderosae Hopkins, *see* Mountain pine beetle
Dendrolasin, in mandibular gland secretions, 99, 104
Diet, relationship between boll weevil pheromone production and, 54
N,N-Diethyl-*m*-toluamide, *see* Deet
3,3-Dimethyl-$\Delta^{1,\alpha}$-cyclohexaneacetaldehyde as boll weevil attractant, 50-54
cis-3,3-Dimethyl-$\Delta^{1,\beta}$-cyclohexaneethanol as boll weevil attractant, 48-49, 51-52
3,6,8-Dodecatrien-1-ol, termite trail pheromone from decayed wood, 66
cis-7-Dodecen-1-ol acetate, sex pheromone of cabbage looper moth, 7-8, 12
trans-7-Dodecen-1-ol acetate, sex pheromone of false codling moth, 9, 11, 12
Dufour's gland, 63, 81, 99

E

Epoxides, as sex pheromones for
the codling moth, 13
Ethyl dihydrochrysanthemumate,
attractant for rhinoceros
beetle, 150
Eugenol, as attractant, 148, 150, 152
European chafer, attractant for,
148, 149
Exocrine glands, 62, 69, 96

F

Frontalin, 27-28, 32-33
Fall armyworm sex pheromone, 6, 8,
11, 12-13
False codling moth, sex pheromone
of, 9
Farnesene and related compounds in
mandibular gland secretions,
75, 99, 104
Feeding stimulants for boll weevil, 42
Fire ant, *see* Imported fire ant
Frass as source of pheromones, 23,
29, 46-47, 51

G

Geraniol
with eugenol as Japanese beetle
attractant, 152
in Nassanoff gland secretions, 73-74
Glomerin [1,2-dimethyl-4(3)-quin-
azolinone], 117
Gypsy moth, 10, 11, 13

H

"Hair pencils" of male butterflies, 6
2-Heptanone
as alarm pheromone, 70, 77
multiple functions of, 87
as trail pheromone, 67, 68, 83
3-Heptanone as alarm substance, 78
Heptyl butyrate as yellow jacket
attractant, 149
trans-10, *cis*-12-Hexadecadien-1-ol,
see Bombykol
cis-7-Hexadecen-1-ol acetate,
see Hexalure

2,4-Hexadien-1-ol butyrate as yellow
jacket attractant, 149
Hexalure, as pink bollworm sex
attractant, 12, 14
2-Hexenal from oak leaves
masking of, 15
as trigger for release of sex pheromone, 6
Honeybee, 67, 68, 79-80
multiple functions of pheromones
of, 84-86
Nassanoff gland secretions of, 88-89
Hydrocarbons
in defensive secretions, 99, 119
in mandibular gland secretions, 75, 83
Hydrogen cyanide, 114, 135, 136
Hydrogenation, in identification, 30,
36, 47, 49, 50
9-Hydroxy-2-decenoic acid, effects on
honeybees, 73, 85-86

I

Imported fire ant (*Solenopsis saevissima*)
multiplicity of action of pheromones, 86
necrophoric behavior, 76
trail pheromone, 63-65
specificity of, 81
Infrared spectra
of cue-lure, effect of deuteration, 154
use in identification, 26, 30, 46, 48
Insect populations, monitoring of, 4,
147-149
Interspecies responses, 24, 30, 35, 68,
75, 77-84
Ips confusus (LeConte), 22, 56
compounds of attractant pheromone of, 23-27
Iridodial, 101-104
Isolation of attractants, 26, 29, 32, 34,
35-36, 44-47
2-Isopropenyl-l-methylcyclobutane-
ethanol, 48, 51-53

J

Japanese beetle attractants, 150, 152

Synthesis *(continued)*
  of *Trogoderma inclusum* attractants,
    36
  of *cis*-verbenol, 26-27

T

Termites, 65-66, 69
trans-3, *cis*-5-Tetradecadienoic acid,
  see Megatomoic acid
cis-9-Tetradecen-1-ol acetate,
  pheromone of fall armyworm,
    8, 11
cis-11-Tetradecen-1-ol acetate,
  see RiBLuRe
Trails, persistence of, 65
Traps, use of attractants in, 14, 43,
    147-149, 151-152
Trigona spp., *see* Stingless bees
Trimedlure, 148, 150, 151
Trogoderma inclusum LeConte, attractant

pheromones of, 22, 35-36

U

Ultraviolet spectrometry, in attractant
  identification, 26, 30
Unsaturated acids as releasers of
  necrophoric activity, 76

V

Valeric acid as attractant, 37
cis-Verbenol, 23-27
trans-Verbenol and verbenone, 27-29

W

Western pine beetle, 22, 27-32, 152

Y

Yellow jackets, 148-149